GW01271103

Akshat Rathi is an award-w
News. He is the host of *Z*
Green. He has a PhD in orga
Oxford, and a BTech in chen
Chemical Technology in Mumbai. He has worked for *Quartz*, *The Economist* and the *Royal Society of Chemistry*. His writings have also been published in *Nature*, *The Hindu*, *Guardian*, *Ars Technica*, and *Chemistry World*, among others. He lives with his wife in London, UK.

Praise for *Climate Capitalism*:

'Climate innovation has accelerated far faster than many realize and by shining a spotlight on the solutions and innovators driving progress, *Climate Capitalism* is an important read for anyone in need of optimism about our ability to build a clean energy future'
Bill Gates

'Addressing climate change will make us richer, happier, healthier, more equal and more safe. Do we take the bargain? That is the animating question of Rathi's illuminating and incisive book, which offers the dazzling and deeply reported argument that the answer should be, overwhelmingly, yes'
David Wallace-Wells,
author of *The Uninhabitable Earth*

'It's easy to feel fear or despair in the face of humanity's greatest challenge, but fortunately work on solutions began decades ago . . . Rathi's brilliantly written account of some of those stories is an inspiration to keep going in a fight which we have no other option than to win'
Bryony Worthington, member of the UK's House of Lords

'Few books on either climate or capitalism manage to be as insightful as they are readable, but Rathi cracks it. He delivers his powerful and hopeful message with both substance and style, and reminds us of the immensely important role of great storytelling as we reimagine our economy'

Paul Polman, co-author of *Net Positive* and former CEO of Unilever

'There are very few people as well-situated as Akshat Rathi is to describe and assess our current efforts to cope with climate change . . . An inspiring book!'

Kim Stanley Robinson

'Bold . . . Climate change is a crisis that requires urgent action, but Rathi shows how we can harness capitalism to tackle it. Give it to the doomsayer in your life'

John Schwartz, professor at UT Austin and former *New York Times* reporter

'Are you suffering from climate anxiety? Go take a few deep breaths and then pre-order this book. You'll learn about fascinating people who show that solutions for climate change are both possible and profitable'

Will Mathis, reporter for *Bloomberg News*

Climate Capitalism

Climate Capitalism

Winning the Global Race to Zero Emissions

AKSHAT RATHI

JOHN MURRAY

About the cover

The long number indicates the amount of carbon dioxide in the atmosphere in October 2023, about 417 parts per million. That's 50% higher than pre-industrial levels. The expiry date 12/50 denotes the deadline of 2050 by which the world should get to net zero carbon dioxide emissions to meet global climate goals.

First published in Great Britain in 2023 by John Murray (Publishers)

This paperback edition published 2024

1

A CIP catalogue record for this title is available from the British Library

Paperback ISBN 978 1 529 32994 0
ebook ISBN 978 1 529 32996 4

Typeset in Bembo MT by Hewer Text UK Ltd, Edinburgh
Printed and bound in Great Britain by Clays Ltd, Elcograf S.p.A.

John Murray policy is to use papers that are natural, renewable and recyclable products and made from wood grown in sustainable forests. The logging and manufacturing processes are expected to conform to the environmental regulations of the country of origin.

Carmelite House
50 Victoria Embankment
London EC4Y 0DZ

www.johnmurraypress.co.uk

John Murray Press, part of Hodder & Stoughton Limited
An Hachette UK company

Contents

I

The Framework

It's now cheaper to save the world than destroy it.*

The first time that became clear to me was in 2016. Donald Trump was campaigning to be the US president and, though he didn't really care about fighting climate change, he was making pronouncements about 'clean coal' technology. I was a science journalist at the time, writing about everything from the birth of stars to the manipulation of atoms. My editor asked me to find out if clean coal was something worth writing about.

What I learned was that Trump mistakenly thought the term referred to coal that could be somehow cleaned after it was mined. Instead, there was a technology called carbon capture and storage, which could trap and bury the CO_2 produced when coal is burned. Crucially, I found start-ups with technology to draw down already existing carbon dioxide from the air, which opened the potential to not just slow down climate change but even reverse it. What started out as a single article about a flawed marketing slogan aimed at making a political point turned into a year-long investigation into breakthroughs in a climate technology.[1]

That series of stories gave me a new perspective on the climate problem. In every sector of the economy – from renewable power

---

* That may have always been true, if we were to accurately value all the things that help humans thrive on this planet. But what's remarkable is that it's true even when we value things in the limited ways in which the capitalistic system does.

to green cement and laboratory-grown meat to electric airplanes – a race was under way to find solutions to cutting global emissions. Politicians, bankers and techies were finding more to agree on with climate activists than they ever had before. Something had shifted and the world hadn't taken notice just yet. The decades-long case for reducing emissions – because it is squarely in the self-interest of humanity – had finally sunk in, and we had entered a phase of focusing on the solutions.

That conviction is now plain to see. In the last few years the US has passed the world's largest climate bill, the EU has enshrined in law its Green Deal, and India and China have set net zero goals, alongside all the major economies. These era-defining moves have happened even as the world deals with the economic shocks of a once-in-a-century pandemic and an energy crisis spurred by Russia's war on Ukraine.

How did it all happen under capitalism? The extractive economic system built over the past few centuries is set up to maximize profits and tends to concentrate wealth in the hands of the rich. Some have argued for a long time that the capitalist pursuit of endless growth at all costs is the main cause pushing the planet to the brink.

There is no denying that unfettered capitalism has contributed to warming the planet. It had become clear decades ago that polluting the commons of our atmosphere would come at a cost. Polluting for free was always going to be a limited privilege. Not pricing in that 'negative externality' – as economists put it – has been the greatest market failure of all time.[2]

But even the most ardent opponent of capitalism, Noam Chomsky, isn't convinced that it would be possible to replace the system with one that's better for the environment in the little time available to deploy solutions globally. There is 'no conceivable possibility' of overthrowing capitalism and managing 'the kind of social change that they're talking about within the time scale that's necessary to solve' the problem of climate change, he says.[3]

If the world's average temperature rose to 2 °C rather than 1.5 °C above pre-industrial levels then the global economy could be $100 trillion poorer.[4] Reaching the more ambitious target would require reaching zero CO_2 emissions by 2050. That's less than three decades to rebuild the energy system that underpins modern civilization, rethink the agricultural system that feeds 8 billion of us, and reinvent the relationship humans have with the planet itself.[5]

It cannot be addressed by the same form of uncontrolled capitalism that is partially responsible for the excess greenhouse gases stuck in the atmosphere. At the same time, reforming capitalism might be the only practical way to get to zero emissions so quickly. This book shows why it's possible to harness the forces of capitalism to tackle the climate problem – and how the work has already begun.

Let's start at the beginning, somewhere in Africa 200,000 years ago. From there the newly evolved *Homo sapiens* slowly spread across the world.[6] Ingenuity helped humans develop ways to grow food, control fire, and use new materials, from wood and stone to bronze and iron. We grew in number, from a few thousand in small tribes to a civilization of hundreds of millions by the eighteenth century, with the ability to travel en masse not just over land but also the oceans.

Then we shifted gear. Our ability to harness fossil fuels at scale accelerated progress like nothing before. With access to cheap, abundant and reliable energy, the human population quickly grew tenfold. On a 10,000-year timeline, the population growth curve follows the shape of an ice-hockey stick: a gentle increase followed by an almost vertical rise. Fossil fuels didn't just give us energy; they opened new worlds. Coal powered both trains and electric turbines, bringing modernity. Oil fuelled cars, airplanes and rockets to further shrink the world. Natural gas made the fertilizer needed to feed billions and turn previously inhospitable places into metropolises.

Much of this progress was unequal until the 1980s, as global inequality increased between countries during the preceding two centuries mainly because of colonialism. I was born in 1987 in Nashik, a small city near Mumbai. Soon after, India and China opened their economies to global trade. Just as capitalism and fossil fuels had done for rich countries, now they powered growth in two of the most populous nations and pulled hundreds of millions out of poverty.

My family experienced social mobility in ways that my ancestors could never have imagined. For most of his life my grandfather, who never studied beyond high school, worked on a factory floor dyeing clothes. My college-educated mother and father launched their own business. As India became richer, incomes grew and opportunities expanded. When my soon-to-be dad flew to Europe on business in 1985, some twenty members of my extended family travelled for six hours to the airport to watch him take off. There's a photo of him at the airport with a garland around his neck and a sheepish smile on his face. By the 2000s my parents had earned enough to build a home for the family and support my sister and me in getting the best education possible.

All the while another hockey-stick-like phenomenon was playing out in the background: the concentration of carbon dioxide in the atmosphere was rapidly rising. The last time it had reached such levels was more than 800,000 years ago, when sea levels were as much as 25 metres higher and *Homo sapiens* had yet to evolve.[7] The unchecked use of fossil fuels, which enables humans to have longer, healthier and wealthier lives, has unleashed climate instabilities that threaten the very fabric of life on Earth. It's possible that entire swaths of the planet will become uninhabitable under business-as-usual scenarios.

The ferocity of the climate impacts we are witnessing around the world – from epic fires in California to apocalyptic floods in China and Pakistan, just in the northern hemisphere summer of 2022 – still seems to take us by surprise. Some say that it shouldn't, because the scientific work behind understanding climate change goes back a long time. A brief look at history reveals a more complex story.

Our understanding of the greenhouse effect dates back more than 120 years, when it was shown that certain gases in the atmosphere create a blanket around the Earth, trapping the sun's heat and raising the planet's temperature.[8] By the 1960s a rough understanding of the theory of climate change was in place with a link to ever-increasing burning of fossil fuels. But scientists couldn't be certain of the impacts. If there were to be any disasters arising from heating up the planet, they remained decades away. Any environmental concerns that existed at the time were squarely focused on polluted water and filthy air.

The oil crises of the 1970s revealed for the first time how addicted humans were becoming to the ever-growing use of dirty and often imported fossil fuels. It forced some to focus on efficiency measures, performing the same activities while consuming less energy. That led to the development of fuel-thrift cars, more electricity-efficient appliances and better insulated buildings. It also spurred a new generation of scientists and entrepreneurs to search for alternative sources of energy: the Americans got working on solar power and lithium-ion batteries, the Danes on wind power, the French on nuclear power, and so on – technologies that have proved crucial to slowing down climate change, even though global warming was not why they were developed.

After much geopolitical wrangling between Western oil consumers and Middle Eastern oil producers, in the 1980s access to fossil fuels was secured and the focus on efficiency programmes and alternative sources became less important. It proved to be bad timing. This was the decade when scientists began to harden their view that unchecked warming could be devastating, not beneficial, as some living in cold, northern countries had previously thought. Even global leaders such as Britain's Prime Minister Margaret Thatcher and US presidential candidate George H. W.

Bush had started raising their voices on the global stage that something needed to be done about the problem.[9]

In 1990, when I was just a toddler, a United Nations-backed body of scientists at the Intergovernmental Panel on Climate Change (IPCC) published their first-ever report on the state of climate science.[10] In a unique procedure, more than 150 countries endorsed the report's summary that burning fossil fuels was increasing the concentration of greenhouse gases in the atmosphere and that the phenomenon would lead to an increase in global average temperatures. Given the strong scientific case and seemingly wide political buy-in, large-scale efforts to cut carbon pollution should have started in the 1990s. But they did not.

That's because, unlike in the 1970s, when it was in the interests of the fossil fuel industries to work on clean energy sources, the glut of oil that followed in the 1980s gave the producers all the incentives to maintain the status quo. A group of American and European oil companies, including the likes of Exxon and Shell, handsomely funded a disinformation campaign to sow doubt about climate science and delay stricter regulations that would otherwise have become inevitable.[11]

That campaign worked particularly well in the United States, which even today remains the largest cumulative emitter of greenhouse gases. Fossil fuel interests captured the political machinery of one of the two major parties (and the grip has only tightened since). The lack of US support meant that most global meetings on climate typically ended in little real action, if not outright failure.

Take the example of the Kyoto Protocol. In 1997 more than eighty countries signed a pact that legally bound rich countries to a reduction in their greenhouse gas emissions.[12] That same year, however, the US Senate passed a resolution – with ninety-five votes in favour and none against – to ensure the country would not agree to any such mandatory limits.[13] Unsurprisingly, the Protocol, which came into effect in 2005 without US participation, did not lead to the outcome the world was hoping for. Emissions continued to rise. And the pace picked up, especially after China joined the World Trade Organization in 2001.

That sped up globalization like never before, with China becoming the factory of the world. And its growth, of course, relied on fossil fuels – specifically coal. The country's annual emissions grew threefold in a decade. The scale of the operation is explained by this statistic: between 2011 and 2013, China used as much cement as the US did in all of the twentieth century.[14] Cement is one of the most polluting products humans make at scale, with annual production accounting for nearly 8% of global carbon dioxide emissions. The rise of China, and later India, as fossil fuel users is one reason why half of all greenhouse gases released since the Industrial Revolution have been emitted in the last thirty years.*

With increased warming, as predicted, came more intense climate impacts. By 2015 the urgency helped create a coalition of the willing among global leadership. Finally, after years of negotiations at United Nations climate summits, 195 countries signed the Paris Agreement. It set a target of keeping global average temperatures well below 2 °C relative to pre-industrial levels, and trying to limit it to 1.5 °C.[15]

Even though the Paris climate agreement was made up of voluntary commitments to cut emissions, it was the first-ever global treaty aimed at keeping catastrophic climate change at bay. Global corporations and financial markets took that as a signal that, eventually, these commitments would lead to national regulations. In turn, that made the financial case to act stronger than ever. Fortunately, decades of investments in green technologies

---

* Though China and India are now the largest and third-largest annual emitters respectively, the US and Europe still hold the title of largest cumulative emitters.

had already begun to make them cheaper than their fossil-fuel-burning alternatives.

In 2017 Trump pulled the United States out of the agreement and reversed many of the domestic regulations that were helping his country cut emissions. But that couldn't subdue the forces that had begun to act. If anything, the urgency to act on climate became a key part of toppling Trump from power. And then it helped Canadian Prime Minister Justin Trudeau get re-elected, gave Germany's Greens their strongest-ever mandate and ousted Australia's climate laggard Prime Minister Scott Morrison.

While these political upheavals occurred, the Covid-19 pandemic caused millions of deaths and plunged the world into a severe economic recession. And still, many countries committed vast sums of money to green activities, hoping that it would lead to faster recovery.

Russia's attack on Ukraine in February 2022 further complicated the recovery. It caused Europe to drastically cut imports that fuelled President Vladimir Putin's war machine. In the short term the sanctions that followed caused fossil fuel prices to spike globally and forced countries to burn whatever they could afford, regardless of the greenhouse gas impact. But in the long term they reinforced the case that distributed clean energy resources aren't just crucial to tackling climate change; they are now central to greater energy security.[16]

Crucially, public concern over climate risks remains at its highest levels. That's visible not just at the ballot box but also on the streets and our social media feeds. Young people, who began protesting on the streets in 2019, are back at it after the pandemic lockdowns ended.

Capitalists have also woken up to both the cost of inaction and the opportunity of action. Private capital is being redeployed, with more than $35 trillion worth of assets – more than a third of all invested assets – now aligned with global environmental, social and governance goals. It's already having an impact.[17]

Before the Paris Agreement was signed, the world was on track to warm by at least 4 °C compared with pre-industrial levels. That would have made entire regions of the world uninhabitable, forced hundreds of millions of people to migrate and reversed much of the progress made in the last 200 years. Since then, the world has corrected course. The worst-case scenario is now 3 °C of warming and, if the current net-zero pledges are met, we are on course to meet the less ambitious Paris goal of keeping warming below 2 °C. But that's still not enough to avoid some of the worst impacts.

We now live in a two-track world. As long as we keep emitting billions of tons of carbon dioxide, the world will see continued climate extremes leading to the loss of lives and livelihoods. That things keep getting worse, however, should not hide the fact that the scale of climate action is also growing. There are now more people working on solutions, more money available to scale those solutions and more government policies in place to help reach emission goals.

Climate Capitalism is about how we tackle climate change within the world's dominant economic system and ensure that the wheels of progress don't come to a halt or, worse, go into reverse. I won't give you one solution or one route that will get us out of this mess, because that's impossible. Instead, my goal is to give you a framework for understanding how we got here, what tools we have at hand and how we are already using some of them to ensure that future generations can also see their lives improve.

The framework relies on three major actors – technology, policy and people – that are continually shaped by money, power and politics. Each chapter uses a successful example to understand how the climate solutions set can be built while progressing the global priorities of economy, security and welfare.

In Chapters 2 and 3 I will show how the Chinese have used capitalism in their peculiar way to become the world's largest maker and buyer of electric cars and batteries, giving an insight into the playbook China has used to create a commanding lead on almost every green technology. Then in Chapter 4 we'll turn to India, a country with a population that's now greater than that of China, but that's much further behind in its development story. Unlike all prior major economies, the success story of solar power in India shows how developing countries with messy democracies and weak governance can still grab on to the opportunity to leapfrog the fossil fuel era and start building clean energy at scale.

The lessons India has to share are crucial for other developing countries, who must adopt them in a bid to speed up the energy transition. But how does that happen? Chapter 5 looks at how international institutions such as the International Energy Agency (IEA) play a vital and little celebrated role in influencing those necessary changes.

In Chapters 6 and 7 we'll turn to the United States, the world's largest historical emitter and home to the greatest number of billionaires. Through the story of Bill Gates, who is one of the biggest private funders of climate technologies and whose lobbying efforts have been crucial in landing the biggest US climate bill, we'll understand how private capital and government regulations can work in tandem. We will then look at the limitations of US government policy to learn how crucial technologies such as carbon capture and storage failed to take off and what can be done to make a necessary technology work.

Fixing capitalism will mean not just reforming how business is done but also completely transforming some industries. Hardest will be the change-over of oil and gas companies. In Chapters 8 and 9 we'll examine two completely different attempts and learn how government policy plays a crucial role in enabling businesses to clean up. As the fight to cut emissions matures, so does the legal frameworks that help speed it up. In Chapter 10 we'll see how to make good climate laws and how that can transform countries and businesses. And, finally, if laws don't go far enough, in Chapter 11 we'll look at examples of how corporate shareholders are starting to use their power to force businesses to change.

Humanity has gone through major energy transitions in the past. The transition from wood to coal kickstarted the first industrial revolution. The move from coal to oil in the early twentieth century launched the second industrial revolution. We have now entered the third major transition, as the world turns away from its dependence on fossil fuels and towards clean energy. None of us can insulate ourselves from the changes coming our way.

Trying to meet climate goals is going to reshape our civilization. This book is about an age that will be defined by the race to zero emissions. Rewiring the global economic system will entail fundamental changes to everything, from how we live and how we travel to what we eat and wear. In short, how we exist.

It has taken a combination of government policy and private capital to scale technologies and create institutions that are finally starting to bend the emissions curve in the right direction. As the world moves away from fossil fuels, the impacts won't be limited to the companies responsible for extracting that carbon. Entire sectors of the global economy – transport, utilities, heating, cooling, chemicals and agriculture – that are reliant on fossil fuels will have to readjust to using clean alternatives.

One of the core tenets of capitalism is the creation of a marketplace of ideas. In a world faced with uncertainty, competition should allow for only the best ideas to succeed. The most passionate capitalists fear that climate action will bring in government intervention that would kill the market, as it tries to redirect the economy. That is not an unreasonable fear.

But what is different today than when Adam Smith birthed capitalism is that, for the first time, humanity has a plan for how to transform at least one major part of the economy – the energy system – over the next few decades. It's a plan backed by decades of science and it has received the backing of every country on the planet. Executing on that plan will require moving in certain set directions, but governments can and must enable those changes without killing competition.

Climate Capitalism is an antidote to the dominant narrative that because we've ignored the climate crisis for so long, it will soon be too late. While it's true that we've not done enough yet, we're nowhere close to being too late. Regardless of the arbitrary warming thresholds set in the Paris Agreement, the science is clear that avoiding every bit of warming is beneficial. It will also be cheaper to achieve climate goals – tens of trillions of dollars cheaper – than to deal with the costs that come from the damages caused by missing them.

These stories of extraordinary individuals and powerful forces will help you see the world in a different, and perhaps more optimistic, light. Above all, I hope to give you the ability to distinguish between the solutions that make a meaningful impact and those that are fanciful distractions.

Let's start in the world's largest current emitter and second-largest economy: China. Even though it's led by the Communist Party, the country's rise over the past three decades is the result of allowing capitalism to flourish in a controlled manner. There's no better way to understand that massive transformation than taking a look at China's playbook in creating the world's largest market for electric cars.

2

The Bureaucrat

Wan Gang cuts a diminutive figure, but when he speaks all ten people sitting around the table listen intently. In an opulent Shanghai hotel conference room lit by golden chandeliers, he is surrounded by executives from international car giants including General Motors (GM), Ford, Peugeot, Nissan, Honda and Tesla, and leaders of Chinese car companies like Geely, Chang'an and SAIC. It is the eighth annual China Auto Forum, in April 2019, and merely three months after the US electric car company Tesla began construction of a factory in Shanghai. The focus is on the transformation of an industry that is turning towards electrification. The executives are aware that what Wan says here can change the fortunes of their companies.

Many have tried to create a mass market for electric cars over the past 140 years, but all have failed. The widely held belief is that if anyone can succeed, it will be Elon Musk, the eccentric, ambitious and obscenely wealthy CEO of Tesla. But when the history of electric vehicles is written, it might be Wan Gang who will stand tallest.

The Musk-Tesla story is lore. Founded in 2003 by Martin Eberhard and Marc Tarpenning in Silicon Valley, Tesla struggled to get off the ground. Elon Musk, who had become wealthy on the back of start-ups like PayPal, began investing in the company in 2004 and took an active role in product design. After Eberhard was ousted following internal conflicts, Musk took over as CEO in 2008 just after the company began selling its first model, the

Roadster. Tesla sold about 2,500 units of the electric sports car, but Musk's stated goal was to make a mass-market electric vehicle (EV). With every iteration, the car models got cheaper and sales grew – turning Tesla into the world's most recognizable electric car brand and the world's most valuable car maker. As of 2022, the company was selling more than 1 million cars annually.[1] Still, the cheapest Model 3 – one that Musk promised would be the affordable car – costs well above $35,000 (nearly £28,000).

Wan Gang's story is mostly unknown. His rise in the EV world started at about the same time as Musk's. In 2007 the car engineer by training was appointed China's minister of science and technology. In the country's top-down economic system, Wan's policies incentivized the creation of hundreds of Chinese companies tied to making electric vehicles. The country now sells more than 6 million EVs each year.[2] That includes not just expensive cars but the complete range, with the cheapest selling for less than $10,000.[3] Wan's policies have also created some of the world's largest and most valuable companies selling electric vehicles and lithium-ion batteries. And the choices he has influenced haven't only affected already established Chinese car companies; all big car manufacturers in the world – for whom the largest market globally remains China – have been affected.

While Musk fought Wall Street's scepticism and benefited from waves of government subsidies to keep Tesla afloat through turbulent periods, Wan has shown how policy done right can drive technological disruption not just in China but worldwide. Both men are at the forefront of the global project to propel the world from the current economic age into the next – yet it is the lesser known of the two who has had the bigger impact.

In the mid 1960s teenager Wan found himself in the middle of a violent disruption of Chinese society. Mao Zedong's Cultural Revolution pitted rich against poor and urban elites against rural commoners. The Red Guard, a paramilitary force controlled by

Mao, subjected those in the higher classes of society, such as Wan's family, to humiliation, beatings and persecution. The Communist Party shuttered universities and sent students to villages for 're-education'. That's how Wan, a city kid from Shanghai, found himself in Dongguo, a village in Jilin province near the North Korean border, working with other city teenagers to build basic infrastructure.

His work ethic caught the attention of local Party members, and in 1974 he was unanimously elected as a team leader. Worried that because his parents were counter-revolutionaries he shouldn't have been promoted, Wan spoke to the head of the local Party branch. 'Keep at it,' he recalled being told. 'One day your parents will be heroes again.'[4]

After Mao's death, in 1976, universities were reopened and Wan studied physics at Northeast Forestry University in Harbin and then mechanical engineering at Tongji University in Shanghai, one of China's most prestigious educational institutions. He excelled there and won a scholarship from the World Bank to pursue a PhD in Germany. For his doctorate at the Clausthal University of Technology, he studied ways to reduce the noise made by internal combustion engines – the type of engine that powers all fossil fuel vehicles in the world.

In hindsight, the decision to study cutting-edge automotive engineering in Germany was perfectly timed. Following the oil crises of the 1970s, the global car industry was undergoing a period of major change. The German car industry wanted to stay ahead of growing competition from the US and Japan, and was crying out for engineers like Wan.

He received job offers from six car companies, from Volkswagen to Mercedes. In 1991 he chose to join Audi, the smallest of the German majors at the time, reasoning that it presented him with the greatest opportunity to rise through the ranks.

Wan began in Audi's car development division, helping to solve technical issues in design and manufacturing. After five years he realized that in order to climb the corporate ladder at Audi, engineers had to show success in more than one department. He duly moved to production, where he focused on car paint and was soon made head of a division with more than 2,000 employees. To effectively manage them all, he deployed techniques he had learned during his years in Dongguo. On an employee's birthday, for example, he would carry two bottles of beer to the workshop floor and spend time getting to know them. The effort paid off, and Audi eventually promoted him to its central planning division, giving him oversight of a manufacturing process that produced a car every sixty seconds.

During his time in Germany, Wan kept a keen eye on his home country. Deng Xiaoping, who took over as the country's leader after Mao's death in 1976, called the Cultural Revolution a 'grave blunder'.[5] In the late 1980s he set about reforming China's economy, including the country's almost non-existent car industry. He welcomed foreign companies, for example Germany's Volkswagen and France's Peugeot and Citroën, to build factories in joint ventures with domestic players. If foreign companies were worried that their Chinese partners would steal their technology, it seemed like a cost worth paying for access to the country's vast untapped market.

By the 1990s the Audi brand had become a favourite of China's elite; government officials were often seen being chauffeured around in black Audi saloons. As one of Audi's top Chinese-born executives, Wan led many company visits to China, at a time when the country's car industry was expanding.

On these visits he noticed how the industry's rapid growth was increasing air pollution and exacerbating China's reliance on oil imports. If his home country was to go the way of its Western counterparts, as its leaders hoped, then these problems would become intractable. At the beginning of the twenty-first century China was consuming one barrel of oil per person per year, whereas in Germany the figure was twelve and in the US it was twenty.

Wan wanted his fellow Chinese to have the quality of life he enjoyed as an immigrant in Germany but, given China's large population, he realized that it might not be possible. It was quite likely that the country couldn't afford the bill from importing all the oil, even if that much oil could be extracted somewhere, which itself wasn't guaranteed. Fossil fuels are finite. The way out was to develop cars that could be powered by something other than oil.

In 2000 Wan got a chance to share his ideas with Chinese government leaders. Zhu Lilan, the country's science minister at the time, visited Audi's headquarters and factory in Ingolstadt, Germany. During the trip – designed to showcase what state-of-the-art car makers look like – he proposed to her that, rather than continuing to tinker with the internal combustion engine, China could leapfrog the West by using a completely different technology.

At the time, the US produced some 15 million cars each year while China produced only 700,000. But international car companies, such as BMW, General Motors and Toyota, were starting to work on electric cars – powered by batteries or hydrogen – that produced no particulate pollution and reduced the amount of greenhouse gas emissions. And Wan was convinced that this form of transport would be the future of the passenger car. If China were to become a leader in electric cars within the next decade or two, Wan told Zhu, the country could become the electric car hub of the world.

Zhu invited Wan to come back to China and make his case to the State Council, the country's highest ruling body. Wan knew that if he succeeded then he could alter China's history. He found support from Li Lanqing, then vice premier of China and who, in 1952, had started China's first major home-grown car maker, First Automobile Works (FAW).[6] Chinese cities were starting to struggle with the problem of smog. But more importantly, if Wan was right, China could become a technology leader and avoid the humiliation of having to rely on Western countries to bring modernity to its people.

A few months later Wan moved back to China. Under the auspices of Tongji University, which gave him a professorship, he began working as the lead scientist on a secret government programme for advanced vehicle technologies.[7] Along the way he played a key role in convincing important members of the State Council to set up policies that would encourage the development of alternative fuel transport, and, in 2009, he launched the new energy vehicle (NEV) programme that would reshape China's car industry.

Wan's political acumen was essential. 'The automobile's importance to growth, trade, innovation, military technology, and the environment is, for practical purposes, immeasurable. The industry is a point of national pride,' wrote Levi Tillemann in *The Great Race* in 2015. 'Since the time of Henry Ford, no automobile industry in the world has ever become internationally competitive without that kind of government intervention.'[8]

In the 1930s the US government paid for the construction of more than 100,000 miles of roads under President Franklin D. Roosevelt's New Deal. It later set up research programmes to push for more fuel-efficient engines and established improved safety regulations. In the same decade the Japanese government provided cheap loans to domestic car makers, funded technology programmes and undermined US players through tariffs to protect domestic companies. In other words, China's industrial policy approach, which would rely on subsidies and regulations, was a tried-and-tested method to boost the car industry.

Wan's plans were bigger still. The car makers he would unleash wouldn't just serve Chinese customers but would make the sorts of cars that would dominate the future of the car industry – by throwing away internal combustion engines and placing all the country's bets on zero-emission transportation.

Electric cars aren't new. In fact, in the early twentieth century there were more of them on the roads than there were internal combustion engine cars. On dung-strewn streets dominated by horse-drawn carriages, petrol-powered cars emitted a new fetid exhaust. They required a hand crank to start the engine, an inconvenience that could also cause injuries because of mechanical kickbacks. By contrast, electric motors powered by lead-acid batteries were a refreshing change: they were started with the push of a button, made little noise, contributed to a smoother ride and didn't add to the foul smell.

Ironically, batteries helped bring an end to the electric car's reign, as manufacturers realized they could put battery-powered electric starters in petrol cars, thereby making hand cranks redundant. A bigger deal was Henry Ford's invention of the modern production line, which drastically brought down the cost of buying a car. At the same time, petrol became even more affordable when the Standard Oil monopoly ended. Governments also continued to rapidly expand road networks and the number of fuelling stations, which allowed car owners to drive longer distances. Battery-powered cars couldn't compete against all those forces.

Until the oil crises of the 1970s it seemed that nothing could stop the rise of the internal combustion engine, but there was a growing fear among fossil fuel corporations that oil might run out. In 1973 the Fourth Arab–Israeli War broke out. In response to US support of Israel, the Arab members of the Organization of the Petroleum Exporting Countries (OPEC) issued an oil embargo, causing a fifth of all gas stations in the US to run dry and plunging the global economy into recession. The big Western fossil fuel companies concentrated their efforts on researching new energy sources, such as nuclear power, and the infrastructure to support them, like lithium-based batteries and electric motors.

Then, in 1979, the Iranian Revolution caused another oil crisis and global recession. Major oil companies were forced into cost-cutting mode, and research divisions were often the first line to get the axe. The work on batteries moved to government-funded laboratories and university departments. Unfortunately for the electric car, the 1980s saw an easing of tensions with OPEC and the oil glut return.

A third attempt to build electric cars came in the 1990s, when California passed regulations to clean its smog-filled cities and created a low-emissions vehicle programme that spurred the development of battery-powered cars. Car makers would be required to meet progressively lower emissions for the fleet of vehicles sold in the country's richest state. The emissions targets were strict enough that simply selling hybrid vehicles may not have been enough. At least partially in reaction to the new California standards, in 1995 General Motors launched the EV1, a two-door coupé that could travel up to 100 miles on a single charge and had a top speed of 80 miles per hour.

Other companies came up with their own versions, but the EV1 stood out – some 800 were leased: more than any of its competitors. It also had the most spectacular demise, after GM recalled the cars and crushed them for recycling. The story was chronicled by Chris Paine in 2006 in *Who Killed the Electric Car?* The documentary alleged that a conspiracy between car makers, oil companies and the US federal government had put the electric car to bed. GM insisted that it was proud of the work it had done on the EV1, spending about $1 billion, and that the technology had found its way into other parts of the business. The main reason for ending the programme was that there wasn't enough demand, which meant existing EV1s could not be repaired and their safety could not be guaranteed.[9] Regardless of who was right, the videos of the cars being crushed caused huge reputational damage for GM and gave electric-car enthusiasts plenty of fodder to keep the conspiracy alive.

That wasn't the end of the road for EVs. In the twenty-first century entrepreneurs such as Elon Musk at Tesla tried to bring electric cars back to the market. As in the 1990s, the roads of California were the battleground, but this time the stakes were higher. The state government's new zero-emissions vehicle programme was promoting electric cars not just to help cut air pollution but also to help fight climate change.

EVs are a climate solution because they are much more efficient than their fossil fuel cousins. For every unit of energy an EV consumes, it can go three times the distance of a similar diesel-powered car. That happens because most of the energy produced when burning fossil fuels is lost as heat, and only a fraction is converted to motion. Electric motors convert more than 90% of the energy stored in batteries into motion. At such an efficiency, EVs produce fewer CO_2 emissions than their fossil fuel alternative, even if the electricity they consume comes from a 100% coal-powered grid.[10]

Just as Musk became the most recognizable name behind the new EV efforts in the United States, Wan Gang was making a case to be recognized as the most influential figure in shaping the future of the global car industry and as the lead character in a far bigger story playing out on the other side of the Pacific Ocean.

Wan's appointment as China's minister of science and technology came one year before China was due to hold the 2008 Olympics in Beijing. An image-conscious Communist Party spared no expense to show off what it was capable of. This would be the first 'green' Olympics, the Party declared, as it announced the closure of coal-fired power stations and factories for weeks, returning blue skies to the smog-choked capital. It also promised to plant enough trees to offset the emissions caused by athletes' air travel.

Wan had been on a deadline ever since being put in charge of China's advanced vehicle programme, back in 2000: to produce electric buses and cars in time for the 2008 Olympics. It wasn't the first time electric vehicles had been launched at an Olympic Games. BMW had produced two prototype lead-acid battery-powered electric cars for the 1972 Games in Munich. But China's plan was far more ambitious; to have 1,000 electric buses and cars ready for the Beijing Games.

By 2007 Wan Gang had many research institutes and industrial partners, including state-owned car makers BAIC, SAIC, Dongfeng and Chery, working on the project. However, China still hadn't mastered the technologies required to make effective electric vehicles: efficient motors powered by advanced batteries and controlled by sophisticated software. Though it had produced and even successfully tested prototypes, China did not possess the manufacturing capability to make 1,000 such vehicles. Rather than admit defeat, the government scaled back its ambitions; a BAIC subsidiary would produce fifty electric buses and Chery would make fifty hybrid electric cars.

Chery had to hire Ricardo, a UK engineering consultancy, to help meet the deadline, according to Levi Tillemann's research. After many long hours the new team had developed a system that could be bolted on to the Chery A5, a compact car, that allowed it to automatically switch between a petrol-powered engine and an electric motor. However, work on the computer algorithms that enabled the switching had begun late. That meant, rather than handing the car over to any driver, Chery had to specifically train drivers for the hybrids who could manually switch between electric and internal combustion engine modes. The BAIC buses seemed to work well, but were retired within three years because their batteries quickly degraded.

None of this came out during the Olympics, and the spectacle had the world enthralled. 'Blockbuster,' wrote the *New York Times*. 'Astonishing,' wrote the *Guardian*. 'The world may never witness a ceremony of the magnitude and ingenuity as that,' said the *Sydney Morning Herald*.

After the Olympians went home, industries restarted and restrictions on car use were lifted. Unsurprisingly, smog returned to Beijing. Within months, in 2009, China overtook the US as the world's largest market for cars, selling 13 million gas-guzzlers. That meant belching out even more particulate pollution – tiny particles capable of entering the human bloodstream and leading to breathing problems. The pollution can cause cancer or stroke, and the higher the number of particles belched out, the greater the harm caused. The Chinese leadership could see the problem from the windows of its Beijing offices. That is why, even though China's EV industry was clearly lagging, the government's support for Wan's ideas to electrify transport did not wane.

Despite the disappointing delivery of EVs at the Beijing Olympics, Wan was able to get approval for a bigger roll-out of new energy vehicles (NEVs) with a hefty subsidy for each new car purchased. The bet was technology neutral, encouraging car makers to make battery-powered cars (BEV), plug-in hybrids (PHEV, large battery and a combustion engine), and fuel-cell cars (FCEV, consuming hydrogen fuel to produce only water as exhaust).

The programme aimed to sell 1,000 NEVs in each of the ten largest Chinese cities by 2012, and the government was prepared to provide as much as $10,000 (£8,000) per car in direct subsidies to incentivize people to buy them. It would also give indirect subsidies to car companies and battery makers in the form of tax cuts and cheap land for factories. The government bill for all that ran into the billions of dollars.

With continued support, the plan eventually began to work. BYD, a Shenzhen-based battery company, launched the plug-in hybrid F3DM – it looked like a carbon copy of the Toyota Corolla – months after the 2008 Olympics. Thanks to the subsidies, there were 10,000 of them on China's roads by 2011.

Even as EVs began appearing on the streets of Chinese cities, the number of fossil fuel cars sold in China continued to increase. In 2012 the country sold 15 million passenger cars. Predictably, pollution worsened, and the figures were available for all to see with the government beginning to openly share air quality data.

Publication of these figures was a surprise. It would almost certainly make the government look bad. But it was a calculated move. In 2014 China's Premier Li Keqiang used the data as the basis for a declaration of war against pollution at the annual gathering of the National People's Congress.

The government had provided a carrot, in the form of direct and indirect subsidies for EV makers. Now it had a stick. Wan Gang's ministry was directed to work with local governments to introduce regulations to control the number of new cars on the roads each year. If a city resident wanted a licence plate for a fossil fuel car, he or she needed to either enter a lottery or bid in an auction. Sometimes the amount they would have to pay for the new licence was higher than the cost of the car itself. For NEVs, it was first come, first served.

In 2011 the country sold about 1,000 BEVs and PHEVs. In 2022 that number stood at nearly 7 million and China had become the world's biggest market for EVs.[11] Some years, the annual rate of growth stood at 300%. As a fraction of all cars sold, EVs now make up more than 25% of total sales – a figure that is already higher than the government target of 20% for 2025 sales.[12] It's clear the future of cars in China is electric, and the country's push has accelerated the electrification of transport globally.

In 2018 Wang Zhigang succeeded Wan Gang as minister of science and technology. Since then Wan has remained a key player in the country's electrification efforts, but his impact was clear even before he left his government job. Between 2009 and 2017 the Chinese government spent over $60 billion on electric cars, according to a study from the Centre for Strategic and International Studies.[13] To put that figure in perspective, it is more than the market cap of General Motors, which produces some 8 million cars each year.

Wan's push also created industrial jewels such as BYD, the world's largest maker of electric vehicles, which counts Warren Buffett as one of its biggest shareholders. It doesn't just sell electric cars around the world; it also sells electric buses.[14] It operates electric bus factories in California and Ontario that have the capacity to build more than 1,000 buses each year.

In that sense, the money the Communist Party spent has already paid dividends. Today, China doesn't just have factories that can produce electric cars; it has an entire supply chain, from the globally mined metals that are used to make batteries to the complex software installed in electric cars. Crucially, the country also has people that can run every level of the supply chain. Though most of this talent is domestic, many Chinese electric car firms are now wealthy enough to poach staff from international companies.

Other countries are trying to play catch-up. Under the Inflation Reduction Act of 2022 – the largest injection of cash from a US government in climate-oriented investments – there is some $100 billion worth of incentives for electrification of transport in the US.[15] Similarly, bullish plans for EVs have been hatched in Europe, where strict emissions criteria have forced car makers to pivot to selling only EVs within the next decade.

During Wan's time as China's minister of science and technology, all the countries in the world signed the 2015 Paris Agreement. Electric cars are a crucial climate solution, and China has shown it is possible to scale the technology quickly. That's led to many countries banning the sale of new fossil fuel cars by 2040 or earlier.[16] Markets covering more than 20% of car sales globally now have a mandate to fully phase out internal combustion engine vehicles.[17]

What Wan Gang, with China's backing, has shown is that succeeding in scaling a green technology requires supportive government policies, substantial public and private investment, and empowering entrepreneurs. Done right, it can also give a country a commanding technological lead over the rest of the world. For climate capitalism to work, all three are required to ensure technologies can scale within a few decades to get the world to zero emissions.

The rapid rise has left oil watchers baffled. The OPEC cartel's 2015 World Oil Outlook predicted that there would be only 4.7 million battery-powered EVs in the world in 2040 – about 2% of the global fleet of cars at that point in time. The world crossed that milestone in 2020. OPEC has been forced to drastically change its estimations, with its 2022 World Oil Outlook expecting there to be about 500 million BEVs in 2045.[18] That still might prove to be an underestimate, with BloombergNEF projections putting the number closer to 700 million in 2040 – about 45% of the global passenger vehicle fleet.[19]

Regardless of the rate of growth of EV sales, global car makers see the writing on the wall. Even if Beijing decides to end all subsidies, the electrification of the global fleet of cars is set to continue. Volkswagen has committed to spending $86 billion on launching new models of electric vehicles by 2025.[20] GM will spend $27 billion.[21] Ford $22 billion,[22] Hyundai $17 billion,[23] and so on.

It's been more than a century since the electric car lost out to the internal combustion engine. The first attempt at resurrecting electric cars in the 1970s failed because the technology was in the hands of oil giants, whose primary business was to extend the fossil fuel era. The second attempt in the 1990s failed because the technology was in the hands of legacy companies whose primary business was to extend the internal combustion engine era. It is the third attempt in the 2000s that now seems certain to put electric cars on top. A combination of strong policy support, technology breakthroughs and well-funded start-ups have convinced even legacy companies to switch sides.

None of this would have been possible without the astonishing development of lithium-ion batteries. And at the heart of that development is CATL, founded in 2011 and now the world's largest battery maker with global A-list car makers as clients. Even if Chinese EVs do not become big brands in the West, Chinese batteries are already inside the EVs that Westerners are buying.

3

The Winner

It was an admission of defeat. But you would never know it looking at the mild-mannered smiles that morning. Angela Merkel, the then German chancellor, was standing next to then Chinese Premier Li Keqiang. On a partially cloudy summer morning in Berlin in July 2018, both leaders made small talk, in between posing for the cameras. The publicity event, one of many planned for the day, was highlighting the collaboration between two of the largest economies of the world.

A few feet in front of them, two men in dark suits sat at a desk with identical leather-bound folders open and pens in hand. They looked up towards the cameras, touched their pens to the paper, and waited a few seconds for the perfect photo to be captured. Then, with the blessing of the elders standing behind, Zeng Yuqun, CEO of CATL (Contemporary Amperex Technology Limited), the world's largest battery company, and Wolfgang Tiefensee, a minister for the German state of Thuringia, signed an agreement committing the Chinese manufacturing giant to building Germany's first large electric car battery factory. The moment passed quickly, and few of those present realized the historical importance.[1]

Germany is known as the home of the car industry and with good reason. It is where, in 1879, Karl Benz built and ran one of the first internal combustion engines designed to power a car.[2] Today, it is home to Volkswagen, one of the world's largest car companies, and other brands such as BMW, Audi,

Mercedes-Benz and Porsche, which are recognized globally for their excellence.[3] The car industry accounts for a seventh of Germany's jobs, a fifth of its exports and a third of its research spending.[4]

The agreement was an acknowledgement that the industry, which had been the country's economic backbone, had finally failed. Not because it couldn't make cars that people wanted, but because it hadn't developed a crucial technology – lithium-ion batteries – that would power them in the twenty-first century.

Batteries are an enabling technology. Apart from powering cars, they can store solar and wind power to be used when the sun isn't shining and the wind isn't blowing. Some entrepreneurs have even built battery-powered airplanes designed for short-haul flights, and there are battery-powered ferries in European waters.[5] As the world races towards its zero-emissions goal, so batteries are proving to be a crucial technology to master. That is why Anja Karliczek, who was in Merkel's cabinet as minister for education and research, called batteries an 'existential' matter for Germany and, more broadly, Europe.[6]

While countries are finally seriously trying to catch up, China has taken a commanding lead. By 2025 China's battery production capacity will be three times as much as the rest of the world combined, according to BloombergNEF estimates.[7]

It wasn't just the Europeans that missed the boat. Even as recently as the late 1990s or early 2000s, few were sure that batteries could do so much and at such low costs. China's rise as the global leader of lithium-ion batteries is now a matter of regret for the oil industry, which invented them; for the Americans, who nurtured the technology towards commercialization; and for the Japanese, who were the first to scale up the technology.

The first thing I noticed when I walked into CATL vice chairman Huang Shilin's twentieth-floor office was the view. It was a grey November afternoon in 2018, and a thick fog rolled over the mountains in front of us, revealing a bay that opens into the East China Sea. For a moment, I forgot we were in the middle of an industrial park.

But reality came rushing back when I walked up to the window and peered down at the construction sites dotted around the tower. Before I could ask, Huang, Zeng's second-in-command at CATL then and one of China's richest men, started describing who was building what in this suburb of Ningde, a growing city of some 3 million people in Fujian province. It is home to CATL's headquarters and some of the world's largest battery assembly factories.

'SAIC is building a factory there,' he said, referring to the Shanghai Automotive Industry Corporation, one of China's largest makers of electric vehicles and partner to two of the world's largest car makers, Volkswagen and General Motors. The car maker would, of course, be acquiring batteries from CATL. 'And that's where CATL is expanding,' he said, pointing to the roofless buildings across the bay. Many have been surprised at CATL's rapid growth. Even Huang admitted that he's barely been able to keep up with demand.

We admired the view for a little bit longer, but I was eager to ask him questions. I hadn't been told how much time I would get with Huang, who was giving his first-ever interview to a Western journalist. We settled on black leather sofas in a corner of his spacious office. The only noise in the room came from an electric kettle that turned on every ten minutes to keep the water at a set temperature. Huang was wearing black trousers and a crisp maroon shirt under a dark blue windbreaker – a lot more casual than I expected for a top executive. At fifty-two years of age, he had a small paunch and a receding hairline, but what grabbed my attention most was his big smile. He handed me a cup of hot water from the kettle, a common practice across China when serving someone water, and we talked batteries – a 200-year-old invention.

★

A battery, technically, is any device that converts chemical energy to electrical energy. The first one, invented by Italian chemist Alessandro Volta in 1799, was simply two metal plates, one made of copper and one of zinc, with a piece of saltwater-soaked cardboard placed in between. Volta's creation, which became known as the Voltaic pile, was revolutionary. It was the first human-made device to provide a steady source of electricity, and its use led to the discovery of many elements on the periodic table and the rules of electromagnetism.[8]

The battery's basic structure hasn't changed since its invention. Just like the Voltaic pile, your smartphone battery has two electrodes (anode and cathode) dipped in a conductive solution (electrolyte) with a separator that stops the two electrodes from coming into contact and causing a short circuit. But the Voltaic pile had to undergo some major improvements before it could become practical for everyday use.

Volta's version of a battery produced only small amounts of electricity while consuming heaps of metal. For instance, to power my flat in London on the Voltaic pile, I would need a combined 11,000 kg of copper and zinc every single day.[9] To make any real use of electrical power, the world had to wait until French physicist Gaston Planté invented the lead-acid battery, sixty years after Volta's invention. The Voltaic pile was a so-called primary battery, which means once the copper and zinc were consumed, the battery had to be discarded. Planté invented the world's first secondary battery. Once consumed, the battery could be charged back up by feeding it with electricity. By the end of the nineteenth century the lead-acid battery was already in use on a mass scale, including to power early cars.[10]

But the battery's reign as the source of energy to power cars didn't last very long, because lead-acid batteries couldn't compete with the distances that could be covered in a car that burned fossil fuel (see Chapter 2). More than 100 years later, lead-acid batteries are still used in cars but only to aid starting the engine, powering the headlights and, more recently, running internal electrical systems like air conditioning and stereo. The rebirth of the electric car had to wait until the invention of lithium-ion batteries (which could store a lot more energy) and a way was found to compete with existing cars.

The development of lithium-ion batteries began during the oil crises of the 1970s, when the big fossil fuel companies were reminded that oil is a finite commodity, and so they doubled down on efforts to find alternatives.[11] One project, helmed by chemist Stanley Whittingham at the US oil giant Exxon, led to the invention of the world's first rechargeable lithium-ion battery: one of its electrodes – the cathode – was titanium sulfide and another – the anode – was lithium metal. But there was a major problem that needed fixing: the battery kept bursting into flames. The risk of fires increases with the amount of energy stored in a small package.

Whittingham was confident the problem could be overcome, and Exxon thought the batteries could eventually be deployed in electric cars. The company approved a budget to address the problems and scale up the project. But before that happened, the 1980s rolled around, the oil glut returned, and Exxon's interest in finding alternatives waned. The company never released a commercially viable battery.

Fortunately, Whittingham's work had sparked broader interest in the field. Over the next decade lithium-ion batteries were the subject of intense work by researchers around the world: in universities, national laboratories, even some companies. Three researchers provided the upgrades that transformed Whittingham's invention into a viable commercial product. First, John Goodenough, then a leading researcher at the University of Oxford, found that using cobalt oxide – instead of titanium sulfide – as the cathode increased how much energy a battery could pack in and also grew the number of charge-discharge cycles it could undergo before degrading. Moroccan scientist

Rachid Yazami found that using graphite (a form of carbon) – instead of lithium metal – as the anode made the battery significantly safer without hugely sacrificing how much energy it stored. About the same time Japanese researcher Akira Yoshino also found that using a carbonaceous anode is better. Finally, Keizaburo Tozawa, then head of Sony's battery division in Japan, and Yoshio Nishi, a senior executive at Sony, put these inventions together. Three scientists, Whittingham, Goodenough and Yoshino, were awarded the 2019 Nobel Prize in chemistry for the invention of the lithium-ion battery.[12]

In 1992 Sony became the first company to commercialize the lithium-ion battery. As an optional upgrade for Sony's Handycam, the battery was 30% smaller and 35% lighter than the standard nickel-cadmium battery. The timing couldn't have been better. As Moore's law observed, the number of transistors on a computer chip tend to double every eighteen months. That led to consumer electronics shrinking, but batteries hadn't caught up until lithium-ion batteries became commercial. It's why Sony's bet proved an instant success. It sold 3 million lithium-ion batteries in 1993 and 15 million in 1994.

Others were quick to jump on Sony's success, including Zeng Yuqun, who, at the age of thirty-one, founded Amperex Technology Limited, or ATL, in 1999. Within two years ATL had produced lithium-ion batteries for 1 million devices and made its name as a reliable supplier. On the back of that success, in 2005 ATL was acquired by TDK, the Japanese firm probably best known for its cassette tapes and recordable CDs.

Zeng and his second in command, Huang, decided to stay on after the acquisition. TDK added Japanese discipline to ATL's manufacturing process and grew its lithium-ion battery business into the newest cash cow: the smartphone market. Soon ATL would go on to supply batteries to both Samsung and Apple, according to Huang.

Starting in 2006, Huang began fielding queries about batteries for electric cars. The earliest request came from Reva, an Indian company. At the time it was making the G-Wiz, a two-seater electric car powered by improved lead-acid batteries. It had a top speed of about 40 kmph (25 mph), a range of 80 km, and took many hours to charge. Reva was looking for a company to supply lithium-ion batteries, which would increase the car's speed and range, and enable faster charging.

The lithium-ion batteries found inside electric cars are quite different from those in portable devices. Vehicles need batteries that can pump out a lot more energy and at a much faster rate than mobile phone batteries. In order to develop a solution, Huang and Zeng created a research department within ATL while simultaneously starting to acquire technology licences from the US that would enable them to build off R & D already happening in the States.

Few Chinese companies at the time were buying up licences or investing millions of dollars into early-stage car battery research and development in this way. Chinese companies have been found stealing or copying from foreign firms.[13] But with its own robust research efforts, ATL broke that mould, and set the stage for Chinese domination of what will be one of the most important sectors of manufacturing in the twenty-first century.

By 2008 ATL already had something to show for its efforts. That year, the Chinese government rolled out a demo fleet of electric buses at the Beijing Olympics (see Chapter 2) – some of which were powered by ATL batteries.[14] The electric bus demo fleet was the start of the government's plan to push for the electrification of transport, a move that would cut deadly particulate pollution and lower greenhouse gas emissions by reducing the number of pollution-belching buses. The Chinese government was under pressure from citizens and the global media to do something about its smoggy skies and to lower its carbon footprint. Huang and Zeng sensed an opportunity. In 2011 they created the spin-out company CATL, C standing for Contemporary to denote their belief that the future of batteries lay in the car business.

About the same time, in a bid to capitalize on a next generation technology, the Chinese government introduced subsidies for electric cars. The catch was that to be eligible the battery had to be Chinese-made. That's when BMW, which was looking to grow its presence in China, partnered with Chinese car maker Brilliance and CATL. In 2013 BMW-Brilliance launched the all-electric Zinoro for the Chinese market. It was based on the design of the X1, BMW's subcompact SUV, and used CATL batteries.

Unlike AA or AAA batteries, which are essentially the same no matter who makes them, electric car batteries need to be custom-made for different car models, to fit the body of the car in the most optimal way possible. That means engineers from the car maker need to work with those from the battery company, exchanging ideas, standards and processes. While working with BMW on Zinoro, CATL added some German engineering skills, such as attention to detail and increasing the reliability of products coming off the factory floor.

'We have learned a lot from BMW, and now we have become one of the top battery manufacturers globally,' Zeng said at an event celebrating Zinoro in 2017.[15] 'The high standards and demands from BMW have helped us to grow fast.' Two years later, CATL would break ground on Germany's first car battery factory, beating the revered German car industry to the punch.

Minutes after I finished speaking with Huang, I got into a light-gold Zinoro waiting outside the building. It took me a few hundred metres down the road to see one of CATL's many battery production lines.

At the entrance of the manufacturing facility, I was told to put on shoe covers to prevent outside dust from entering the factory, and was forced to relinquish my smartphone. No pictures allowed. I was led down a long corridor with spotless white walls that felt more like a hospital than a factory. On one side of the corridor were windows through which we could see the battery production floor, where humans and robots, in equal measure, worked. The humans were covered from head to toe: shower cap, lab coat, gloves and shoe covers. The robots did their jobs naked. The humans spent most of their time observing machines or looking at computer screens. The robots did the heavy lifting and moving of different components between machine lines and elevators.

The lines prepare and assemble the components. The process of making a lithium-ion battery begins by mixing solvents – liquids that can be easily evaporated – with powdered electrode materials, creating a slurry.

CATL primarily uses two types of cathode: lithium iron phosphate (LFP) for electric buses and nickel manganese cobalt oxide (NMC) for electric cars. NMC packs more energy in less space and is thus used in smaller vehicles, but it is more expensive than LFP. In both cases, the anodes are made of graphite (sometimes mixed with a little bit of silicon). The battery gets its name for the charged particle of lithium that shuttles between the two electrodes, as the battery is charged and discharged.

Once the slurries are made, they're painted on sheets of metal: copper for the graphite anode and aluminium for the LFP or NMC cathode. The sheets go through a hot oven, which dries out the solvent and allows the powder to stick to the metal sheet.

In the next steps, nifty machines bring three sheets together: cathode, separator and anode. (Separators are typically high-grade plastics that allow lithium ions to pass through but do not let the two electrodes come into contract.) They are then rolled together and inserted in a casing based on the shape of the battery: cylindrical, pouch or prismatic. Cylindrical cells are used by Tesla and look like AA batteries but larger. Pouch cells are used by Audi and are similar to those flat rectangular batteries found in smartphones. Prismatic cells look like sleek lunch boxes.

The factory I got to see was making prismatic cells, which are used by BMW. Once the rolls are in place, another machine injects the electrolyte into the casing. The battery is now ready to be charged for the very first time.

But the job is still not done. Many prismatic cells need to be connected and linked together to form a battery pack. CATL engineers cannot do that on their own, because each battery pack is custom-made for a specific car model. CATL's engineers need to work with the car maker's engineers and figure out the next steps.

This is where knowledge and skills are exchanged between the two teams. To build a car battery pack for the BMW i3, for instance, some ninety prismatic cells are connected to each other.[16] The batteries are arranged to perfectly fit into the car's chassis. The connected prismatic cells are operated by a battery management system, which can measure many different things about the battery to give detailed information on the health of each battery inside the pack. The system is also equipped to undertake heat management, which kicks in to help batteries charge or discharge at a safe temperature. This level of customization is not cheap, which is why some electric car makers, such as Rivian and Canoo, are developing so-called skateboard platforms, which can be used across different car models.

CATL has become one of the best in the field at managing this complex and delicate production process, and today it has a long list of car makers as customers, from German brands like BMW and Volkswagen to Japanese ones like Honda and Nissan and, of course, numerous Chinese companies too. In just a few short years it has combined Japanese discipline, German engineering and Chinese entrepreneurship to become the manufacturer of some of the world's most coveted electric vehicle batteries.

The battery's importance shows up in CATL's valuation. When it listed on the Shenzhen Stock Exchange in 2018 it instantly minted four billionaires, including Yuqun and Huang. In 2021 CATL's market capitalization was greater than that of Volkswagen.[17] Yuqun's net worth is now greater than Alibaba founder Jack Ma's, who topped the Chinese billionaire leaderboard for many years.[18] But CATL has to keep on its toes to continue to dominate the space, with new start-ups throwing up bold challenges for the future of lithium-ion batteries.

Late in 2020 a Silicon Valley start-up named QuantumScape listed on the New York Stock Exchange. Within weeks its valuation soared beyond that of the US car giant Ford, even though the company had told investors that its next generation battery won't be in a commercial car until at least 2026 and thus it wouldn't have any significant revenue for many years.[19]

Ever since Chinese battery makers took the lead in lithium-ion battery manufacturing, the only way for US companies to make a comeback is if they build a battery that is vastly superior to what's commercially available today. QuantumScape is one of among a dozen other start-ups and large companies betting that solid-state batteries are likely to provide that boost. And QuantumScape's sky-high valuation showed the market was clearly interested.

The inspiration for the leap came from Whittingham's Exxon battery, which used lithium metal on its anode. Lithium is the universe's lightest metal and, for a lithium-based battery, it is the most energy-dense anode material available on the periodic table. But commercial developers had abandoned lithium metal batteries because, even if they didn't catch fire, the batteries degraded quite quickly. Imagine buying an electric car whose driving range falls 20% or more every year.

QuantumScape's big breakthrough was to find a material that would not only make lithium metal batteries safe but also ensure they lasted at least as long as modern lithium-ion batteries did. That material is a solid electrolyte, eliminating the liquid electrolyte in the battery that, researchers had determined, was one of the key reasons why lithium metal batteries were degrading so quickly.

Getting there, however, was not easy. The start-up began work on finding the material in 2011, after it gained a deep-pocketed backer in Volkswagen, which would eventually invest as much as $300 million in the company.[20] Its researchers listed the properties the material needed to have: it must allow lithium ions to flow and resist the degradation of lithium metal. 'We did not know if a material existed in nature that could meet the requirements,' said Jagdeep Singh, CEO of QuantumScape. 'Much less that we would be capable of finding it.'*

When tackling a hard problem, the ultimate weapon for scientists is brute force. That is, conduct as many experiments as possible quickly, learn from the results, and tweak the conditions and perform yet more experiments. It's trial and error, but informed by a huge amount of data and much analysis. For the purpose, QuantumScape had built not just a small army of battery scientists but also artificial intelligence specialists.

To cut down the time needed to go through those iterations, it built a 24/7 operation where researchers took shifts to ensure that all scientific instruments were in optimal use all the time. The results were continually fed into supercomputers that would, like the replicator from *Star Trek*, create new materials from a vast database, but digitally. Those material recipes were then developed in the lab and tested for performance in batteries.

After running 'millions of tests', Singh says, in 2015 the team finally found two materials that could meet their requirements. Then came the next challenge. Though the materials worked in tiny lab-scale batteries, they had to make them at large enough

---

* This quote and all unreferenced quotes are directly sourced from personal interviews.

scale to test in hundreds of large pouch cells that could be put into a car. That work took another five years. 'That may seem like a long time,' said Venkat Viswanathan, a battery expert at the University of Michigan and advisor to QuantumScape, 'but it's how long it takes to solve the "and" problem of battery materials.'

QuantumScape's solid electrolyte didn't just have to allow lithium ions to pass through and reduce the degradation of lithium metal; it also had to be flexible enough to not break inside a battery and easy enough to produce at scale. After running the two competing materials through a series of tests, one emerged successful.

After ten years and spending hundreds of millions of dollars, QuantumScape revealed its breakthrough solid-state battery in 2020. It packed 50% more energy in the same volume as the best commercial lithium-ion battery on the market at the time, and could charge up from 0 to 80% in less than fifteen minutes, which is twice as fast as Tesla's best batteries can do now.

Singh won't give any detail about the solid electrolyte material, apart from saying it's a ceramic. Even photographers have to use filters to avoid leaking the colour, lest that may give competitors a hint. The company has a few more years of development work left to build the large-scale battery lines, and that process is known to throw up a lot of problems.

Though Volkswagen will get first dibs at using QuantumScape cells in a commercial car, the battery company is then open to doing business with any car maker willing to pay. Time is running out, with solid-state battery prototypes emerging from other start-ups like Solid Power and Ilika, and from established companies like Toyota, Samsung and even CATL.

So far, the rapidly growing demand for batteries has caused only a few scares. They've shown in rapid spikes in prices for lithium, cobalt and nickel. Higher prices have induced more companies to bring on new supply for the metals. Beyond the supply-demand dynamics, cobalt specifically has other problems. About a half of the world's cobalt is mined in the Democratic Republic of Congo, which has been criticized for illegally employing children to work in toxic environments.

Lithium-ion battery manufacturing capacity globally is expected to grow to at least ten times current demand in the next two decades. It will require development of new mines in safer locations with stronger environmental and human rights standards. At the same time, the world will also need to build large-scale battery recycling facilities.[21] A Greenpeace study found that some 13 million tons of lithium-ion batteries are expected to go offline between 2021 and 2030.[22]

Studies have shown that about 95% of valuable minerals inside a lithium-ion battery can be recovered. The question yet to be answered is whether it can be done at a price that the industry can afford. That's especially true as the cost of batteries fall. Governments aren't waiting to see whether the economics pan out. The Chinese government has created incentives to build as many as 10,000 recycling facilities across the country.[23]

Battery recycling is even more crucial for Europe. That's because China has a stranglehold on the supply of battery materials. As of 2021, the Asian country accounts for almost all of the world's graphite production, 95% of manganese sulphate production, 80% of cobalt sulphate, 55% of nickel sulphate and almost half of lithium hydroxide.[24] So Europe needs to not just secure access to virgin metals but also to keep them in the loop for longer through recycling.[25] In 2022 the EU set out regulations for the minimum amounts of cobalt, lithium and nickel that must come from recycling.[26] The Inflation Reduction Act in the US is set to spur not just battery factories in the US, but also recycling of battery materials.[27]

Battery-powered portable devices have transformed our lives. Battery-powered electric cars are starting to do the same. And there's a lot more that batteries can disrupt, as safer, more powerful and energy-dense batteries are made more cheaply. That's why Merkel and her ministers consider batteries an 'existential' subject. In 2021 CATL began producing batteries in its new plant in Germany, supplying cells to car makers in the heart of Europe.[28] The focus is not just about supporting Germany's car industry, but about developing an edge on a technology that's going to power the future as we race towards zero emissions. Huang, CATL and China have together fired the opening shot, but there's plenty of room for others to try and catch up.

Between 2010 and 2020 the price of lithium-ion batteries declined 89% – from $1,100 per kWh to $137 per kWh.[29] The next decade is likely to bring another halving, even as pandemic-led supply disruptions cause battery prices to increase in the short term. And because batteries are the most expensive part of an electric car, the steep decline has allowed the likes of Tesla, Volkswagen and Toyota to move electric cars from the luxury section to the mass-market section of their product list.

The decline in battery prices is, in part, because of the work of companies like CATL. Every new low-emissions technology, from solar panels to wind turbines, is expensive at first. Its life starts in the lab, where scientists don't spare expenses to tweak and learn how to get the best prototypes to the factory. Then the technology is transferred to engineers, who break down the steps needed to manufacture many thousands of units. As engineers make more units so they find ways to make production more efficient. That might mean cutting down the amount of waste produced, reducing the time it takes between two steps, or some such detail too mundane for anyone but the engineers to care about. The upshot is that each successive batch can be made that bit more cheaply than the previous one.

If a company has a monopoly on the product, which is rarely the case, then it can make more profit per unit sold. More often, companies in a competitive market pass on some of the savings to customers by lowering the price. But that doesn't necessarily mean lower profits, because lower prices allow consumers to buy more solar panels, wind turbines or batteries – thereby feeding a virtuous cycle.

Technologists chart this process on what is called a learning curve, which has the cost of producing a unit on the y-axis and the number of units produced on the x-axis. For lithium-ion batteries as an industrial product, the learning curve shows that, for every doubling in the number of units manufactured, the cost of producing a unit falls by 18%. This rate is a historical fact, but if the demand for batteries continues to grow then it's also a number that can predict how fast battery cost will fall.

Though quite expensive when first introduced in the 1990s, portable electronic devices powered by lithium-ion batteries proved to be better than those using the nickel-cadmium batteries. Consumers were willing to pay the premium for batteries that were lighter and longer-lasting. As more people bought devices with lithium-ion batteries so the cost of producing the batteries continued to fall. By 2010 the costs were low enough that it became affordable enough to put lithium-ion batteries in early luxury electric cars.

Continuing down the learning curve, some estimate that in the 2020s the cost of lithium-ion batteries will be so low that, on average, it will be more economical to buy and operate an electric car than an internal combustion engine car.[30] At such low costs, lithium-ion batteries are also starting to find applications in other markets.

In September 2017 Hurricane Maria, the world's most intense tropical cyclone, officially killed 3,057 people and plunged the United States territory of Puerto Rico into darkness. It would take the authorities eleven months to restore power across the island, setting yet another record: the longest blackout in US history and the second-longest blackout in the world, ever.[31]

As climate change intensifies extreme weather events across the world so the chances are high that Puerto Rico will soon get hit by another Maria-like storm. Puerto Ricans want to be as prepared as they can be. Within weeks of the disaster some local environmental activists who call themselves 'energy insurgents' began to put in a new system to power at least some parts of the island.[32]

The typical modern electric grid consists of a large power plant – usually burning coal or natural gas – that provides electricity to homes and offices through a network of large and small cables. Break enough of these cables, as Hurricane Maria did, and the whole grid goes down. The new system, called a mini-grid, made use of new and more resilient technologies. The source of electricity generation would be solar panels. But the power of the sun is diffused and distributed, which means you need a lot of solar panels to replace the capacity of a single large power plant. The energy insurgents distributed the panels across the region – some placed atop homes, others raised on unused land.

Fossil fuel power plants have another major advantage: they provide 'dispatchable' power. When you turn on a light switch, at that very instant someone needs to generate and dispatch electricity to satisfy your demand. Fossil fuel power plants can rely on the stored energy from millions of years ago, which can be released in a fire as and when needed. Solar power, on the other hand, is reliant on the diurnal cycle and the fickleness of cloud cover. To make the mini-grid work just as reliably, Puerto Ricans added lithium-ion batteries. When the sun is shining, the solar panels can provide some power for consumption on demand and some for squirrelling away in the batteries. When the sun sets, batteries become power plants, releasing stored energy on demand.

The biggest advantage the new system offers to Puerto Ricans is that if another hurricane strikes, as it surely will one day, and a mini-grid loses connection to the main electric grid, then it will still be able to operate as an energy island. As it currently stands, mini-grid system won't meet all the demand, but most of the basic services will be able to function. That the renewable-powered mini-grids produce no emissions is a great side benefit.

None of this, however, would have been remotely possible without the steady decline in battery costs over the past thirty years. Electricity is one of the cheapest commodities in the world, which means all the equipment to generate and distribute it needs to be low cost as well. The income per capita in Puerto Rico is just a third that of the US average; the fact that many Puerto Ricans can afford to pay for batteries and solar panels to build mini-grids shows just how affordable these technologies have become.

There's still room for the battery costs to fall further, which is great news considering some 700 million people worldwide still lack access to reliable electricity.[33] Just as cheap mobile phones allowed many people in developing countries to bypass the last-century technology of landline phones, so distributed sources of energy in the form of solar panels and batteries could allow many to bypass the need for large electric grids.

And batteries can play a big role in supporting already existing electric grids. If there's a short cut to solving the emissions problems, it is to electrify as much of the world as possible and power that with clean electricity. That's why, as climate goals tighten, many countries and regions are setting ambitious goals for renewable penetration on the grid.

Energy storage is a crucial technology that can enable solar and wind – now the cheapest sources of power – to work more reliably. That is not to say that hydro, nuclear, geothermal or other forms of dispatchable, emissions-free sources of power don't have a role to play, but energy storage can help speed things up, regardless of the clean source of power you choose.

At CATL, Huang was involved in building lithium-ion batteries the size of shipping containers. In a country like China, which has both a growing number of electric cars and huge solar and wind power capacity already installed on the grid, these batteries can serve two purposes. When an electric car needs to recharge, the batteries can provide high-speed charging without adding demand on the grid. When there are no cars charging, the batteries act as storage for the grid. In 2022 Huang left CATL to pursue opportunities in EV charging and energy storage.[34]

Technological progress has certainly made it easier to tackle climate change. But it would be wrong to think that we have all the technologies we need, or even the materials, to build as much of it as is needed to reach zero emissions. A world powered entirely by solar and wind power is not yet within reach. Current technology can take us up to 80% or 90% renewables, but the closer you get to 100% the harder it gets.[35] That's because the longer the amount of electricity that needs to be stored on the grid, the cheaper the storage medium needs to be.

A 2019 study looked at what it would take for renewables to meet the cost competitiveness to provide electricity at all hours of the day, matching a gas power plant. The answer it came up with is that the cost of energy storage needs to fall to as little as $10 per kWh, less than one-tenth of today's lithium-ion battery prices.[36] And none of the experts I spoke to believe that lithium-ion batteries can ever get so cheap, which means energy storage breakthroughs will be needed in the decades to come. There are already start-ups making batteries with iron, an extremely cheap metal, which could meet that ultra-low cost requirement.[37] But those companies will need lots of public and private support to scale.

With China's prowess in green technologies, it will soon reach peak emissions and find a way to get on the path to net-zero emissions by 2060 that the country's leader set in 2020.[38] However, what happens in the country of my birth – India – could alone determine whether or not the world meets its climate goals. India's size and yet-to-come economic growth could mean a lot of fossil fuel use, which is something the country is entitled to as a matter of fairness. Even so, an astonishing story of growth in solar power shows it is taking an alternate route to prosperity than the one used by the US, Europe and others.

4

The Doer

Although it was an early afternoon in January – the middle of an Indian winter – the sun in Pavagada beat down without relief. Sweating uncomfortably, I stood on a small patch of land between the road and the barbed-wire fence protecting the construction site. Behind me, large trucks bearing shipping containers with the logos of global brands like Maersk and Hanjin kicked up dust storms and drowned local farmer Srinivas's voice with a steady groan of engines.

Just three years ago the only way to reach this piece of land would have been to ride a two-wheeler on bumpy dirt roads miles away from the nearest main road. That day, Srinivas and I had been driven here in an air-conditioned car on smooth asphalt roads. The roads had only recently been built – as had one of the world's largest solar farms, which we'd come to visit.

The Pavagada Solar Park, situated a few hundred miles from Bengaluru, the tech capital of India, covers 13,000 acres – approximately the size of 6,000 football pitches – with the capacity to generate 2,000 MW of emissions-free electricity. That's enough electricity to meet the annual needs of an Indian city of 3 million people. Starting in 2016, the country's government began leasing land from farmers, including Srinivas, to build the solar park. Without it, he said, pointing to the construction site where rows of neatly laid solar panels stretched to the horizon, hundreds of farmers in debt with large agricultural loans would have been staring into the abyss.

If it weren't for the lease income, 'I couldn't have sent my girls to university,' Srinivas confided. The fifty-year-old father of three girls wore a crumpled white shirt. His brow, like mine, exuded pearly beads of sweat. He gave a sigh of relief and told me how happy he is that his daughters won't be farmers when they grow up and will get the chance at a better life than he's had.

On the car ride that brought us here, I got a glimpse of the worries that have been mounting for farmers in the region. As we drove from Bengaluru to Pavagada, tall trees and greenery gave way to open land dotted with shrubs and cacti. The southern metropolis of Bengaluru is bestowed with tropical savannah weather that supports the large green areas that are the basis for its nickname: the garden city of India. Though only a short distance away, Pavagada is in a semi-arid zone with little rain and temperatures that range from hot to very hot all year round.

The farmers here are no strangers to drought, but it's become worse over the past two decades.[1] Many local planters have tilled the land for generations, and none have seen droughts with such frequency and lasting such long periods. Farmers have tried to adapt by growing groundnut and yellow lentils, which can make do with the dwindling water supply. Those who can afford it have dug bore wells to use groundwater for irrigation. In the past ten years even that resource has come under threat. Many of the wells have dried up and there doesn't seem to be an easy way to replenish the ancient reserves.

'Sooner or later, climate change will turn the land barren,' said Yethiraj, an advisor with Tumkur Science Center, a local educational charity. Srinivas and Yethiraj, like many people in south India, prefer not to share their full name, which can include the caste they belong to, their father's name and the name of the village their family comes from. It's partly for the practicality of not having to repeat a long name and partly for privacy. The naming convention shows the deep roots people of the region have to the lands they come from, and many people don't want to reveal their full name for fear of giving away their identity.

Millions of farmers around the world today are having their lives upended as climate change renders vast stretches of land as fallow. The decades they spent tending the productive earth are coming to an end. A 2017 study found that nearly 60,000 Indian farmers have killed themselves in the last thirty years because of crop failures linked to climate change.[2] The drying up of farmland around the world is likely to accelerate with every fraction of a degree of global warming, and soon those numbers could be orders of magnitude higher.[3] It is one among thousands of examples where the worsening impacts of climate change are more acutely felt by the poor and most vulnerable.

No single solution can help all farmers. Some, like Srinivas, could benefit from the transition to clean energy. But for those benefits to materialize, new rules have to be put in place to help create a system where government and business can work together to allow developing countries like India to tap the unending supply of solar energy that bathes its people every day.

India neither invented the solar cell – that was the United States – nor manufactures it at scale, as China does. Nonetheless, the solar boom in India that began in 2015 could provide a blueprint for other sun-soaked tropical countries that have the greatest potential to deploy solar.* But they do not have the financial power of rich countries like Japan or Australia. What's happening in India can be a model for countries in Africa, Central and South America and South East Asia, which combine to make up more than half of humanity.

Srinivas has been able to preserve his connection to the land of his ancestors because of a decision made some 5,000 miles away

---

* Tropical countries are also likely to suffer the worst impacts of climate change.

in 2015. That year, 195 countries signed the Paris Agreement. It achieved something that no environmental agreement had achieved before: every country not only acknowledged that climate change is one of the biggest threats facing humanity but also signed off on a target that, if met, would stop some of the worst catastrophes from happening.

The year of the agreement, India set a goal to deploy more than 100,000 MW of solar power by 2022 by way of a declaration of ambition on the global stage. The country then had less than 5,000 MW installed,[4] meaning it was planning to scale up its solar deployment twenty times in less than seven years. Although solar is now the cheapest source of electricity in much of the world, when India set the goal, companies still needed large government subsidies to enable solar to compete with other sources.

In setting a renewables goal then, India was trying to pull its weight in global climate diplomacy, even though its contribution to the problem is small. India's share of the total fossil-fuel-related carbon dioxide dumped into the atmosphere stands at one-eighth that of the US.[5] If India was to hit net-zero emissions by 2050 alongside the US, the country's total contribution to global greenhouse gases would be one-fifth that of the US.[6] These comparisons are for absolute emissions. The figures would skew further in India's favour if per capita emissions are considered, with India's population more than four times the size of the US population.

Each megawatt of solar typically requires more than six acres of land. In a densely populated country, the Indian government has had to buy or lease much of the needed land from private owners or farmers. For hundreds of farmers, like Srinivas, the annual lease payments may not fully compensate for the lost farming income but do provide a decent, steady source of income that eases a challenging life.

Public perception of the Paris Agreement is often that it's just another vague agreement between governments. Some countries have deviated from the promises they made, such as when the US elected a climate change denier as its president and abruptly withdrew from the pact. But the vast majority have remained committed. Crucially, the scientific, economic, social and moral case has only become stronger.

Despite its weaknesses, the Paris deal has created real momentum for climate action, and is reshaping our world. Its greatest achievement has been to send a signal to private companies that governments are finally ready to take on the challenge of climate change and commit to scaling the solutions needed. It is the link between world leaders in Paris and Srinivas in Pavagada.

'That's the beauty of the Paris Agreement,' according to Thomas Spencer, an analyst at the International Energy Agency and former researcher at the Energy and Resources Institute, an Indian think tank.[7] 'Part of its strength is its weakness as well. It appeals to so many different sectors, so many different levers of action and just sets the direction. It doesn't actually say you have to do it.'

An even greater force behind the momentum is reality. Many of the dire predictions scientists have been making for decades are starting to come true. The number of 'natural' disasters is on the rise, with 2020 recording as many as 980 according to reinsurance firm Munich Re.[8] These disasters cost the global economy $210 billion.[9] In the 1980s the number of floods, droughts, wildfires and storms stood at about 300 each year and cost a lot less on average. Climate change has made the impact of these disasters worse than the impact of natural variability alone. Damage from such disasters is likely to continue to grow as long as the world keeps emitting more carbon dioxide, bringing chaos and adding to uncertainty – exactly what governments and businesses fear most.

Crucially, scientists are now able to say with a high degree of certainty whether certain extreme-weather events are linked to human-caused climate change. An extraordinary heatwave struck the west coast of North America in the summer of 2021, causing hundreds of deaths and smashing temperature records set decades before. Within a week scientists published a study showing that the intensity of the event would have been 'virtually impossible' without climate change.[10] When catastrophic floods submerged a third of Pakistan in 2022, the scientific group World Weather Attribution found that climate change 'likely' increased extreme rainfall, which contributed to the flooding.[11]

The worsening droughts in Pavagada have meant an increasing number of farmers are failing to pay off debts or cannot sustain themselves. Often their first resort is to apply for crop failure insurance, which is underwritten by the government. With more and more claims being filed across the country, opportunities like the solar park in Pavagada are a way for the government to decrease its liabilities without abandoning its responsibilities to support the country's farmers.

But those savings alone would not have been enough for a cash-strapped Indian government to lease Srinivas's land and build a solar plant. The government also needed entrepreneurs like Sumant Sinha, whose company was responsible for building about 15% of the Pavagada Solar Park. They bring the scale of private markets and, together with government, make the enterprise of solar power economically sustainable in the long term.

About a decade before the solar park was completed, Sinha created ReNew Power with the backing of capital from the US investment bank Goldman Sachs. Since then, he has built a world-class renewable energy company in a country where it's tough to run a good business, especially one that deploys new technologies.

Goldman Sachs had no previous experience with investments in Indian renewables. In Sinha, however, it saw someone capable of navigating difficult waters. His previous job, as chief operating officer in Suzlon, India's then largest wind energy company, gave him the experience of getting things done in renewables. Prior to that, as chief financial officer of one of India's largest conglomerates, he had finessed his money management skills.[12]

Sinha started by making a simple case. In the 2000s the Indian economy was growing so rapidly – at an average of about 7% annually – that even recovering from the 2008–9 global financial crisis wasn't a tough task. A developing country typically consumes more and more energy as it expands its economy. In India's case, that energy mostly came from increasing the burning of coal and oil. The country's consumption of the dirtiest fossil fuel grew by 70% in the 2000s – an unsustainable pace.

Fortunately, renewables were getting cheaper at a faster rate than had been projected, and the Indian government was keen to be ahead of the curve. The country already relied on imports for much of its oil and gas consumption. And even though it boasted the world's fifth-largest reserves of coal, it couldn't dig up as much as it was burning – forcing up imports of coal too. Finding new sources of energy was crucial for maintaining energy security and lowering its import bill.

That provided solar momentum, but Sinha's bet was still risky. In 2011, when he founded ReNew Power, India ranked 134th on the World Bank's ease of doing business rankings, much worse than China (79th) and not too far behind Russia (123rd).[13] The South Asian giant is infamous for corruption and red tape, and this makes life difficult for entrepreneurs generally but especially so if they have to work closely with state-owned companies.

At the time, in the renewables business, private players would have to rely on government subsidies to compete in the energy market. And in India, subsidies were promised but never guaranteed. Then there was the problem of dealing with state-owned grid operators, who were responsible for transporting the electricity produced by solar or wind farms. The operators would pay the wholesale price, which they would recoup from the retail payments that customers make for the electricity consumed. Except that the grid operators were heavily indebted, partially because of electricity theft and government-mandated freebies. These problems increased the risk for companies like Sinha's to not get the money they were due.

I first met Sinha in September 2019 at the company's headquarters in Gurgaon, just outside New Delhi. Then fifty-four years old, he was softly spoken and only got animated when talking about new business ideas. But his route to entrepreneurship had been circuitous.

As a young man, there was an obvious pathway to success for Sinha: politics. He is the son of Yashwant Sinha, India's former finance minister and foreign minister, and the brother of Jayant Sinha, former deputy finance minister and deputy aviation minister. But politics didn't hold the same allure for the youngest Sinha. After studying engineering at the Indian Institute of Technology in Delhi and finance at Columbia University, in 2008 he got himself a front-row seat in India's renewable energy boom as chief operating officer of Suzlon Energy, which at the time was the country's largest wind turbine manufacturer.

In 2007 Suzlon had reached a market capitalization of $50 billion and stood as a shining example of India's potential to create multinational giants. The company had customers the world over. But the global financial crisis of 2008 took the wind out of Suzlon's turbines, causing its stock price to fall more than 90% and plunging the company into a serious crisis.[14] It was spending way too much without thinking through how it could repay its debt in lean times, and Sinha says his attempts to steady the ship and correct mistakes were met with resistance internally. The problem wasn't with India's renewable energy market, but how Suzlon was spending money to capture it. It was time to captain his own ship, Sinha realized, and he founded ReNew Power in January 2011.

Tragedy struck early. Sinha had hired a small team, built a pitch deck, and spent his life's savings buying options on potential renewable energy projects, but then he suffered a retinal detachment, a serious condition which if not treated immediately could have lead to permanent vision loss. For four weeks following surgery on an eye Sinha was forced to stay in bed, unable to move his head.

Back up on his feet, he hit another barrier. ReNew was looking to raise $60 million. That proved to be too much for venture capitalists, who typically give Indian start-ups hundreds of thousands of dollars or small millions as a starting investment. And it was too little for private equity investors, who invest hundreds of millions of dollars instead.

After facing rejection from more than thirty investors, Sinha landed an interested party in Goldman Sachs. It was a fortunate coincidence. The US investment banking giant was the owner of Horizon Wind Energy, which it sold in 2007 to Portuguese power provider Energias de Portugal for more than $2 billion.[15] Goldman saw potential in the renewable business and it was looking to find avenues to invest in emerging economies – and Sinha showed up just in time.

As a private equity investor looking to invest much more than $60 million, Goldman asked Sinha to provide a bigger plan for the company. After a few months of back and forth, Goldman agreed to take a majority stake in ReNew with an investment of $200 million.* It was, at the time, the biggest single investment in a renewable energy company in India, and a big bet on Sinha, who hadn't yet built a single renewable power plant.

Securing a large sum in equity allowed Sinha to address a major problem for businesses in India: the high cost of debt. Business

---

*Over the years, it would invest a further $270 million, before the company went public.

loans from Indian banks tend to have interest rates higher than 10%, which means businesses have to be able to make enough returns to repay those expensive debts. That's hard for a business like ReNew, which has high risks on both technology and execution.

'Sumant understood, right off the bat, that renewables was as much a finance story as it was a technology story,' said Kanika Chawla, programme manager with UN Energy, who previously worked at an Indian think tank that advised ReNew Power.

Here's one way to think about the cost of capital. Say I wanted to buy a house that costs $1 million, and I have $500,000 to spend on it. The rest would have to come in the form of a mortgage loan from a bank. I would own 50% equity in the house and the bank would own the remaining 50%. Because both the parties share equal risk, a US bank would be ready to lend me that sum for an interest rate of as little as 2% (assuming I have a good credit history).* That means I will have to pay $10,000 in interest every year. But if I'm only able to find $100,000 up front, and the bank has to lend me $900,000, my equity is only 10%. Suddenly, the bank is taking a lot more risk on the property than I am, so it will ask me to pay a higher rate of interest, say 5% – or $45,000 a year. The lesson is that higher equity would allow me to pay less in interest and thus lower the cost of capital.

The same scenario in India, however, will play out differently because Indian banks charge a much higher interest rate – upwards of 10%. That's largely because the country's central bank has decided to keep interest rates higher. There are three reasons why: the rate at which the country's economy is growing (typically two or three times the US growth rate), the country's

---

* This was true during the 2010s when interest rates were at rock bottom. Following the pandemic and energy crunch, they have increased, and the numbers are now higher in the US and higher still in India.

inflation rate (again typically higher than US inflation), and inefficiencies in the economy (such as the rate of defaults or loss of tax revenue).

Equity has a cost too. If it's your own equity then by choosing to invest in your company, you're giving up interest you may get from a savings account or the returns from investing in the stock market. If it's an investor's equity then you have to promise the investor a rate of return that's higher than what the investor could get from a savings account or the stock market. Add the cost of debt and equity to get the total cost of capital.

Beyond the high cost of capital, the cost of doing business in India is higher than in wealthier countries. That's in large part because of weak institutions. Sinha's job is to build renewable energy projects and provide the electricity generated to the operators of the grid, who are called distribution companies or discoms. Although ReNew Power as a company can own the power plant that generates electricity, its customers are India's state-owned discoms, who then supply electricity to homes and businesses.

Discoms in India typically lose 25% of their revenue due to issues like stolen power and government regulations that force them to give free power to, say, farmers.[16] Worst of all for discoms' balance sheets, federal regulations often prohibit state-owned companies from raising prices to make up for losses.

The upshot is that, on average, discoms are two or three months delayed on their payments to ReNew. In turn, ReNew has to hold on to higher cash reserves to ensure the company can always make its interest payments to banks on time. Sinha acknowledges the risk the company faces if its cash flow is cut. 'We've learned to live with the situation,' he says.

ReNew has developed a team specifically charged with ensuring that state discoms pay up. If discoms were run more effectively, ReNew would not have needed to hold on to extra capital or pay teams to collect dues. That would have helped make ReNew more profits or pass on those benefits to customers, further lowering the cost of renewable power.

I experienced the problems with discoms firsthand. In 2019 I went back to my home town of Nashik for my dad's sixtieth birthday. In addition to celebrations, I also wanted to figure out if powering my parents' home with solar was feasible and economical. The answer was a resounding yes, with estimates showing that the upfront cost would be paid for in less than a decade, as energy bills plummet. The panels would have a life of twenty-five years, making all the electricity produced after ten years essentially free.

Securing the panels and getting them installed wasn't much of a hassle either, with many suppliers and installers competing to get the business. However, once everything was in place, getting the discom to connect the system to the grid took months. That step was necessary for my parents to be able to benefit from getting paid for any excess renewable power they didn't use. The solar panels were on the roof in February, just in time for the summer heat that starts in March. Those few months are the only time my parents needed to use energy-hungry air conditioners. But the discom did not approve the application until the end of May, days before the monsoon arrived and rendered the use of air conditioners unnecessary.

If it's not the discoms failing then there's always something else. Indian state-owned banks hold vast amounts of poor-quality debt invested in struggling coal power plants. These so-called 'non-performing assets' have been weighing heavily on the balance sheets of many public-sector banks. In 2020, for every Rs 100 that Indian banks had lent, they were at risk of not being able to recover Rs 12.[17] A few big defaults could kickstart a domino effect that may cause some banks to fail entirely, resulting in a squeeze on the sources of capital available to renewable energy companies.

Those are worries Sinha and other entrepreneurs in developing countries have to live with. And, yet, despite these structural problems, they have been able to succeed. That's because of an unexpectedly fast decline in the cost of solar power, of about 90% between 2009 and 2019, according to BloombergNEF.* Combine that with India's location on the globe – the closer to the equator the straighter the sun's rays and the more electricity solar panels produce – and you've got the recipe for setting world records for low-cost solar. In many cases, it is now even cheaper than electricity from an existing coal power plant. In June 2020 a 2,000 MW solar project won the bid to provide electricity at Rs 2.36 per kWh or $31.40 per MWh – one of the lowest rates in the world.[18]

Farmer Srinivas is not just a beneficiary of the Indian government leasing his land, Sinha's entrepreneurialism and Goldman Sachs's capital; he also benefits from the long story of innovation that has given the world the modern solar cell. This story shows how scaling any technology, but especially clean energy technology, requires persistent efforts from governments, scientists, private companies and entrepreneurs – each of whom need to be personally incentivized to participate in the process.

Though it feels quintessentially of today, the solar cell has a history stretching back nearly two centuries. In 1839 French scientist Edmond Becquerel discovered that two metals in a conducting solution could generate electricity when exposed to light, but he couldn't explain why it was happening. That didn't stop others from tinkering with the idea. In 1883 American inventor Charles Fritts was the first to develop working solar panels, which used

* The price of solar has continued to decline since, with a brief increase following the pandemic tied to supply-chain shocks and inflation.

the element selenium, and install them on a roof in New York City.

It took the genius of Albert Einstein to explain how solar power actually worked. In 1905 he described a phenomenon he called the photoelectric effect: small packets of light called photons have enough energy in them to knock off loose electrons found in the outer shells of certain elements. Throw enough energetic photons at a metallic surface, and it's possible to free enough electrons to generate electricity. Einstein's paper describing the phenomenon won him the Nobel Prize in Physics in 1921.

Armed with this scientific knowledge, companies began exploring whether solar panels made up of photovoltaic cells could become a marketable product. In 1935 the US company Westinghouse concluded that solar panels would need to convert about 25% of the light energy that falls on them into electricity to become a commercially viable product. The efficiency of selenium solar cells stood at 0.5%.

The next major breakthrough came in the 1950s when scientists at Bell Laboratories – a pioneering research institution in New Jersey that also gave the world lasers, transistors and cellphones – figured out that the same material used to make computer microchips could be used to make solar panels. The silicon-based cell they created had an efficiency of 6%.

In 1954 the *New York Times* ran a story under the headline 'Vast Power of the Sun Is Tapped by Battery Using Sand Ingredient'.[19] It noted that the invention 'may mark the beginning of a new era, leading eventually to the realization of one of mankind's most cherished dreams – the harnessing of the almost limitless energy of the sun for the uses of civilization . . . Nothing is consumed or destroyed in the energy conversion process and there are no moving parts, so the solar battery should theoretically last indefinitely.'

With hindsight, we can say that the story got a few facts wrong. For example, this innovation was never a 'battery' technology

– they are called solar panels rather than solar batteries because panels generate their own electricity whereas batteries only provide the power stored in them. And yet it was prescient in articulating the basic phenomenon of how solar cells work, and grasping the potential of what solar could unleash, even though it would take fifty more years for that reality to crystallize.

Silicon is the second most abundant element in the Earth's crust, after oxygen. In fact, combine those two elements and they become silicon dioxide or silica, which is the major component of sand.

While Einstein found that light particles could dislodge electrons from certain materials, they didn't do it as easily as was needed to make a commercial solar cell. In a silicon photovoltaic (PV) cell, scientists at Bell Laboratories were able to tweak the materials to make the dislodgement of electrons that little bit easier.

PV cells consist of two types of silicon. N-type silicon is made by mixing silicon with a tiny amount of phosphorus or arsenic – elements that have one more electron in their outermost shell to silicon's four. The layer then has an excess of electrons. Similarly, p-type silicon is made with minuscule loads of boron or gallium – elements that have three electrons in their outermost shell. That creates an excess of 'holes' or spaces where electrons can be accepted.

The photovoltaic cell is formed when n-type and p-type silicon are laid down in layers. Because n-type has an excess of electrons and p-type has electrons missing, the layering creates a junction with an electric field across it. That electric field provides just the right amount of incentive needed. When light strikes the n-type layer, it dislodges the electrons from the atoms that held them. Then the electric field forces those electrons to flow through an external cable creating electricity, and they end up meeting with a hole in the p-type junction.

So the only moving parts in a solar cell are the electrons. Mechanical parts suffer wear and tear when they are used, but subatomic particles move almost without friction. A silicon PV cell's life is not indefinite, as the *New York Times* story had claimed, because even the most robust chemical structures eventually break down under the relentless rays of the sun. But silicon PV survives for decades without need for repairs. That's what allows companies to provide warranties on solar panels that can be as long as twenty-five years.

Over decades working on this technology, scientists have found that solar cells with multiple P–N junctions – called multi-junction solar cells – are more efficient. That's helped push the efficiency of commercial silicon PV to higher than 20%.[20] In 2020 US scientists at the National Renewable Energy Laboratory created a six-junction solar cell that reached 39.2% efficiency.[21]

Silicon PV has the lion's share of installed solar in the world, but many other materials can be used to produce panels with similar or even higher efficiency. Gallium arsenide (GaAs) solar cells are known to be among the most efficient, but the material is so expensive that it's only used in space applications, such as on satellites or Mars Rovers. Cadmium telluride (CdTe) solar cells use less material and were once cheaper to produce, but they are not as efficient as silicon PV. Most recently, scientists have developed perovskites that could be cheaper than silicone but are yet to catch silicon in terms of efficiency.

The Bell Laboratories solar breakthrough came at a time when the US government was more interested in a different clean energy technology. In the 1950s the Atoms for Peace programme saw annual research spending by the US government on nuclear power reach more than $1 billion, while solar secured merely $100,000. In the 1960s solar found a niche customer in the growing space programme, powering satellites that would not have otherwise found ways to operate for more than a few days or months using energy from on-board batteries or fuel. That meant investments grew into the millions of dollars, sustaining progress.

The 1970s brought difficulties for the oil industry. It was getting harder and harder to find new wells, and market watchers worried that oil could run out. Then, in 1973, the Arab–Israeli War broke out. Oil-exporting Arabian Gulf states cut off supply to countries, such as the US, who supported Israel. Within months, a fifth of all US gas stations didn't have any fuel, and the prices at those that did had skyrocketed.[22] The shock led to resurgent interest in solar from a US government worried about dependence on imported oil and from American fossil fuel companies looking to find a way to make money on other energy sources. Oil and petrol companies, including Exxon, Mobil, Arco and Amoco, built entire divisions dedicated to solar.[23]

But fears over peak oil supply were premature. The Iranian Revolution of 1979 stressed the oil markets once more, and the 1980s brought back a glut that pushed prices down and lowered profits. American oil companies began cutting costs wherever possible; alternative energy research divisions were among the first to go. Many sold their solar portfolios to European oil companies, where the political situation was more favourable than in the US where companies focused a lot more on short-term profitability.

On the other side of the Pacific, the energy-poor country of Japan also took up the solar baton. Its government launched the Sunshine Programme in response to the oil crisis of 1973 and later passed the Alternative Energy Act in 1980.[24] The government used taxes on electricity and coal use to fund much of the research, and it encouraged homegrown companies to invest in the commercialization of solar power. It also launched subsidy support for installing solar on 10,000 roofs during the 1990s. Even though US interest in solar declined, Japan's funding for solar research stayed steady, which allowed Sanyo, Kyocera and Sharp to become leading global producers of solar cells in the early 2000s.[25]

Germany showed similar interest in solar power research after the oil crises of the seventies, but government support for deployment of solar only began at scale in 2000.[26] That's when the country passed legislation to guarantee a premium price for anyone generating solar power and feeding it to the grid.[27] That led to German companies rushing into the solar market. Japanese subsidies for solar could not keep pace, and by 2010 Germany was responsible for half the world's deployment of solar, with companies like Siemens and QCells benefiting from the domestic boom.[28]

Finally, the baton passed to China, which by then had become the world's factory and was seeing an opportunity in solar panel manufacturing. As with Germany and Japan, government support in China was crucial. The vast majority of that support, however, did not come in the form of subsidies for domestic deployment of solar power. Instead, China subsidized solar cell manufacturing explicitly to meet demand abroad.[29] That meant that between 2005 and 2010, 90% of Chinese-made solar panels were exported to countries that were offering generous subsidies, such as Germany, Spain and Italy.

The financial crisis of 2008 caused many countries to start pulling back government support for solar. Though China weathered the financial crisis much better than Western economies, the decline in solar panel demand internationally caused the price of solar panels to halve between 2009 and 2013. To support its domestic industry, which had birthed stars like JinkoSolar, SunTech and LONGi, China created subsidies for domestic solar deployment, such as premium rates for renewable power generators and tax rebates for solar cell manufacturers. China is now both the world's largest maker of solar panels and its largest market.

In 1954 Bell Laboratories put the cost of solar power at about $280 per watt.[30] In 2020 utility scale solar cost about $0.20 per watt – a decline of 99.93% in a period of seven decades.[31] Its learning rate works out at about 20%, which means that for every doubling of installed solar power capacity the price of solar drops about 20%. The history of silicon PV cells shows clearly how important government funding and policies are helping to achieve those cost declines in clean energy technology.

In 2010, under Prime Minister Manmohan Singh, India set the goal of reaching 20,000 MW of installed solar by 2022, generating about 2% of the country's electricity.[32] Just as Germany and Spain were ending the premiums they had previously offered to renewable power generators, Indian discoms began offering them. That encouraged people like Sinha to build companies like ReNew Power to help meet the need. Learning by doing helped Indian companies to lower the cost of solar quite quickly. By 2014, when Narendra Modi took over as prime minister, the cost of solar panels had fallen so low that the new leader of India's government decided to quintuple the country's 2022 target to 100,000 MW.[33]

The economic fallout of the pandemic meant India failed to reach its 2022 goal. It finished the year at a little more than 60,000 MW of installed solar capacity.[34] Still, the figure is three times the initial goal of 20,000 MW. And India has reached it despite not having the financial heft of rich countries or much money in the form of international aid.

Unlike Europe or even China, the Indian government could not afford to sustain the subsidies for very long. It duly dialled back premium electricity tariffs for renewable power suppliers. But to sustain the momentum on renewables, the government introduced a 'reverse auction' system for solar and wind power. In a typical auction, buyers successively bid higher amounts until the highest bidder takes home the painting. In a reverse auction there is only one buyer (the Indian government) and many sellers (renewable developers) that successively bid lower prices for their solar or wind power, and the lowest bidder wins the contract to build the renewable power plant. The competition to win contracts has forced Indian companies to improve efficiencies in their work and supply chain to further lower the cost of renewables. The reverse auction system has been credited with delivering record low bids for solar power, and it has been copied by countries including Brazil and South Africa.

At the same time, as India has increased its share of solar and wind so it has had to find innovative ways to ensure electricity is available to consumers even when the sun doesn't shine and the wind doesn't blow. Working in partnership with companies like ReNew Power, the government has introduced new renewable energy initiatives such as hybrid systems that use both solar panels and wind turbines (often wind speeds peak when the sun is not shining), systems that combine solar or wind power with batteries, and systems that use coal power plants to fill in the gaps when solar or wind cannot.

In 2020 the government asked for bids for the world's first round-the-clock renewable energy auction, and ReNew Power won.[35] The company will have to provide 400 MW of zero-carbon electricity for 80% of any day through a combination of solar, wind, battery storage and hydro projects. To overcome the intermittency of solar and wind, it is likely the company is going to have to build much more than 400 MW of power-generating renewable assets to meet the bid's needs. One approach would be to build 500 MW of wind, 900 MW of solar and 1,600 MWh of batteries, according to BloombergNEF calculations.[36] The good news is ReNew should be able to do all of that for Rs 2.9 per kWh, or $38 per MWh – cheaper than coal power.[37] Few industrial sectors in India have cooperated so closely with government, something that has proven to be a win–win for both.

Through helping farmers like Srinivas, the solar story in India is about climate justice – ensuring that the needs of the poor and the vulnerable are prioritized, because they are the ones who are suffering the consequences of climate change the most. Through overcoming the challenges of doing business in an emerging market with high cost of capital and weak public institutions, as

Sinha has done, it is also about entrepreneurship. None of this would have been possible without external forces, such as the steep learning rate that made solar so cheap. Climate capitalism shows how, time and again, it is a combination of people, policies and technology that is needed in often different amounts to grow the momentum of the global clean energy transition.

India still has plenty of difficulties to overcome in order to meet its ambitious climate goals. Like most countries, it heavily relies on imports of solar panels from China. In 2020 the two countries clashed on disputed borders near India's northern region of Ladakh. The skirmishes gave rise to an anti-China sentiment in India that may lead the government to increase taxes on Chinese imports or even set limits on the amounts that can be imported. That could make solar more expensive in the country, at a time when it needs to further accelerate clean energy deployment.

One solution is for India to build its own solar panel manufacturing base. Entrepreneurs like Sinha will be at the forefront again, but it's going to be a risky investment. Initially the factories won't be able to compete on price against imports from China, which means they'll have to depend on government support that will either provide subsidies to Indian manufacturers or add import taxes on Chinese panels. That support will also have to be maintained for a few years so that entrepreneurs can recoup sunk capital. The government-industry coordination that has created the solar boom in India will have to remain steady through the next phase of growth.

Barring any major mishaps, however, India's solar future is bright. In 2021 Modi upped the country's solar goal one more time, to 280,000 MW by 2030.[38] For context, that is six times as much renewable energy capacity as the UK has installed as of 2020. Meeting the ambitious goal will inevitably need many more people like Sinha and Srinivas to play their part. India is showing the world how to lay down new rules that allow governments and industry to work together to take advantage of technology trends and provide clean energy for growing needs. Instead of going through the transition that the richer countries of the world did, from coal to gas to green electricity, India is in a position to skip that middle step – and help build a greener future for itself and others.

What happens with solar in India could be a lesson for so many other developing countries. But that doesn't guarantee adoption. The problems those countries face are worse: even higher cost of capital, rampant corruption and a lack of skills across business and government to scale clean energy. Bodies like the International Energy Agency play a crucial role in the global system by helping countries move to cleaner sources while ensuring there's enough energy available for citizens and businesses.

5

The Fixer

The Paris Agreement, signed by every nation on the planet in December 2015, proved to be a turning point for the climate fight. But it almost did not happen. One of those who made it possible was the energy analyst Fatih Birol.

On 13 November of that year, terrorists in Paris killed 130 people in bombings and shootings. It was the bloodiest day in France since the Second World War, and it was only two weeks before the start of UN climate talks where 40,000 attendees were expected.[1] Crucially, it happened a mere three days before a high-level meeting of energy ministers.

Birol, a Turkish economist, was then newly appointed head of the International Energy Agency, and he was going to lead the meeting of the ministers. He planned to make the case to all nations that the solution to climate change lay in how we use energy, which causes more than three-quarters of all planet-warming emissions. And he hoped to broker an agreement on moving to clean energy among the ministers from twenty-nine countries, which made up nearly 40% of all global energy consumption. If he succeeded, it would provide the crucial momentum needed at the climate talks to come in the following days.

Even if the terrorist attack hadn't happened, it would not have been easy to get countries like France, which had very few coal power plants, and Turkey, which heavily relied on coal, to agree on ways to reduce energy emissions. But if there was any chance of reaching an agreement, it would necessitate getting all the ministers in one room to thrash out the details.

After the attack the French government declared a state of emergency. That meant any meeting of international delegates would require a lot more security at a time when the country's security apparatus was being stretched to its limits. The meeting seemed like it might not happen.

But Birol worked closely with the US to get France the security assistance it needed and convinced the French government of the importance of the meeting, despite the horrific photos plastered across newspaper front pages. It would be one among many high-profile, behind-the-scenes diplomatic wins Birol would score as he worked to put the might of the IEA behind transitioning to cleaner sources of energy.

'Without fixing the problems of the energy sector, the world has no chance whatsoever to solve the climate problem,' Birol said. 'I have a big responsibility as the head of the IEA to lead the global energy system in a sustainable way. This is the one and only reason I'm doing this job.'

For most of its existence, few people knew about the IEA outside the energy industry and high-level politicians. The agency's main focus is publishing heaps of reports – often many in a month – that lay out the state of affairs affecting global energy security. The technical reports, produced by the IEA's army of number crunchers, feed into government decisions that mostly don't make the news headlines but keep the world churning.

But Birol's proclamation that the IEA would now support the energy transition was a big turnaround for an organization steeped in fossil fuels. For the first four decades of the IEA's existence since being founded in 1974, the focus of its reports was carbon-based fuels, especially oil. The fuel drove (and often can still drive) global economies and big-game geopolitics. The IEA's role was to serve its oil-consuming members, mainly Western economies, in tackling the ups and downs of that market.[2]

In the 2000s, as the world began to seriously consider how to wean itself off those dirty fuels, there didn't seem to be any international organization capable of doing what the IEA did for the fossil fuel era. Thus in 2009 the United Nations created the International Renewable Energy Agency (IRENA) to publish reports on new, clean sources of energy and set up global meetings that would push countries to lower emissions.

As Birol pitched for the top job at the IEA, in 2014, he realized that the agency's mandate would have to expand if its pronouncements about the energy world were to remain influential. Solar and wind power prices were falling more rapidly than many experts had predicted, including those at the IEA. The era of cheap batteries and green hydrogen was on the horizon.

IRENA's reports were a welcome addition, but Birol realized that they weren't providing the systemic view an agency looking at the entire energy sector could. In any case, the two institutions weren't on an equal footing. IRENA had been born inside the UN, which meant its membership included all countries in the world. That gave it the power of having a bigger base of popular support when it came to finding a consensus, but the bureaucracy needed to secure agreements among so many members typically meant acting more slowly. In the 1970s IEA burst out of a cauldron of crisis directed by a few of the richest and most powerful countries to help them not be crushed by oil markets that had been hijacked by the Arab countries. Ever since, its mandate of energy security has tended to rank higher on most countries' priority lists than the decadal threat of climate change. That's allowed the IEA to build close ties with old, powerful energy ministries among its member countries, while IRENA works alongside newer, less powerful ministries created to support clean energy and the environment.

Birol also spotted an opportunity in the limitations of the UN system. He had seen how the UN's annual climate gathering, called the Conference of the Parties, was often too cumbersome to make real progress. Any COP deal needed a consensus among all the nearly 200 countries in the world. Perhaps the IEA could fill a gap by getting its few but powerful members to come to an agreement first, which could ease the process at the COP meetings later.

The IEA shows how international institutions can rebuild to become fit for the climate era, and why many must go through such a transformation to remain relevant in the twenty-first century. Many intergovernmental institutions will play a major role in accelerating, or hindering, the transition. That's because, even though every country needs to reach net-zero emissions within decades to avoid climate disaster, each has different strengths and weaknesses and thus will need support from across borders to meet that goal.

The IEA's origin story continues to shape it even today. The 1960s saw a rapid increase in the use of cheap oil across Western countries. Cars in the US, for example, grew larger and heavier, reducing their fuel mileage. Nearly two-thirds of that increased oil demand was met by the Arab members of OPEC. It gave those countries extraordinary power to make use of what came to be known as the oil weapon, which was deployed, to great effect, in 1973, when a group of Middle Eastern countries that were members of OPEC stopped exporting oil to a handful of European countries and the US. The embargo came in retaliation for those nations' support of Israel in a regional war against Arab countries. It caused oil prices to spike and created a real risk of the commodity running out altogether in the West. That could have led to public protests and huge political backlash.

Western powers recognized that such a weapon could exist only because of coordinated action among OPEC members. The only way to make it ineffective would be to have a coordinated response. They had to get creative to build a new international organization that could quickly ready such a response. In 1974 the IEA was born under the existing framework of the Organisation for Economic Co-operation and Development (OECD) – whose membership is comprised of mostly rich Western countries – thereby avoiding the need for another international treaty to be signed. The IEA's immediate mission was to ensure oil-consuming Western countries could always have access to the fossil fuel and at reasonable prices.

The organization proved crucial in creating 'strategic petroleum reserves' – enough oil stored within borders to last up to ninety days – in many IEA countries. When periods of chaos inevitably came, such as the Gulf War in 1990, investing in the IEA paid off for members as they saw a damping down of volatility in the oil markets and a lowering of the negative impact on oil-consuming countries' economies.[3]

Such a coordinated response required a freer exchange of energy-related information with the IEA. That's how the agency built up its number-crunching prowess, as it carefully assessed how members were using energy and from where the fuels to provide that energy were being sourced.

This unique expertise enabled the IEA to start creating scenarios of what the future of energy could look like. And as it grew its membership over the decades that followed so it developed its data-hoovering capacity and ability to forecast the next stages of the global energy markets. But the twenty-first century brought in new, decidedly different energy challenges, and that future-gazing power started fading. Indeed, when the UN created IRENA, in 2009, it was as a counter-force to the fossil fuel bias of the IEA.

It took another five years for IEA members to finally wake up to a new global energy reality: when Russian troops invaded Ukraine and annexed Crimea in 2014, it caused panic among European countries that received a large share of Russian natural gas through pipes in Ukraine.[4] That led the Western powers in the IEA to finally approve an expansion of the organization's energy security mandate to include not only global gas markets but also a recognition of the changing nature of energy to add threats to the power sector including those wrought by climate change.

It was also when the IEA's member nations promoted Birol, then the agency's chief economist, to lead the organization into the modern era. Given that the agency's previous six chiefs were all government bigwigs, elevating the chief economist to the top job may have seemed like electing the maths club president to prom king. But the logic was sound.

A degree in power engineering and a doctorate in energy economics landed Birol a job at OPEC as an oil market analyst at the start of his career, in 1989.[5] The IEA hired him in 1995. He would go on to play a pivotal role in developing the organization's marquee product, the annual *World Energy Outlook* (*WEO*), which uses exclusive access data secured directly from governments to project energy supply and demand for years to come. The two skills he needed were facility with numbers to explain what the complex models say and the ability to build relationships across geographies and cultures to get diplomats on his side.

'He's one of the best communications experts I've ever worked with,' says Brian Motherway, head of energy efficiency at the IEA. This isn't just praise you'd shower because Birol's your boss. Few people I've seen have the capacity to hold the attention of an audience and make sense of numbers around dry and abstract energy statistics with as much ease as Birol does.

Birol has white hair and black eyebrows. Although he has spent nearly three decades living in Paris, he still has a Turkish lilt to his English. He doesn't look or sound like the typical Western energy expert. But he uses his not-part-of-the-club demeanour to full advantage. For example, he throws into conversations his love for the Turkish football club Galatasaray or some Turkish saying, to bring to life drab details about energy. It's his way of saying, I may not know your culture, but there are enough things in common between us to have a dialogue. As US Energy Secretary Ernest Moniz put it, Birol is 'very sensible in how he listens, how he talks to people, gets help, gets allies, builds up consensus to allow him to step out.'

While the rhetoric matters, it's nothing without substance. The strength of its data is what gives IEA its 'backbone', Birol says, and what enables the organization to have productive relationships and wield influence globally. 'Friendship is not only when they like you – as we say in Turkish – because of your beautiful eyes,' he says. 'But because you take the right position and you have a backbone.' It is these friendships – built on numbers and facts – that have helped Birol undertake the most consequential transformation of the IEA: from caring only about carbon underground to beginning to care about carbon everywhere.

Historically, new IEA chiefs travel to OECD capitals to make sleepy inaugural speeches to kick off their terms. Instead, Birol started his tenure by flying to Beijing and New Delhi – the capitals of non-IEA nations. He understood that the organization's transformation had to begin with expanding its membership. When he took over the agency in 2015, the twenty-nine member states overseeing the IEA made up 38% of world energy use. It was missing two of the top three largest energy users – China and India.

Eight years after Birol vowed to build 'a truly International Energy Agency' to his first audience in Beijing, the agency has thirty-one traditional members and four countries in the wings waiting to become full members. It also has eleven 'association countries' that pay much smaller dues and don't get the same voting power as members, but who do get to participate in IEA meetings and tap into its data expertise.[6] Bringing China, India, Brazil and Indonesia – countries that face the dual challenge of meeting increasing energy demand while tackling emissions – under the IEA's umbrella lets Birol boast that the organization now represents the interests of those consuming more than 80% of global energy.

Birol takes pride in the IEA being nimble in its decision-making and more hands-on with its member countries. It works closely with governments, academia and the private sector, then moves that knowledge from places that have it to places that need it. Motherway agrees, noting that 'We know people in ministries as individuals.'

Take the example of hydrogen. In June 2019 the IEA concluded a year-long project with a special report on hydrogen that declared the clean burning fuel was finally 'on a path to fulfil its longstanding potential as a clean energy solution' and called on governments to take 'ambitious and real-world actions now'.

Less than a year later, as governments around the world began to consider stimulus measures to revive pandemic-hit economies, they turned to the IEA for advice. The upshot: the EU unveiled a hydrogen strategy to create an industry with €140 billion in revenues by 2030,[7] Germany announced €9 billion in hydrogen funding,[8] and Joe Biden's climate plan called for 'carbon-free hydrogen' to be cheaper than hydrogen made from fossil fuels.[9]

It's not all because of the IEA's special report, of course, but the agency's fingerprints are there to see. Timur Gul, the IEA's head of energy technology policy who also led the work on hydrogen, described to me how ideas initially laid out in the agency's technical reports filter through to decision-makers, and sometimes get converted into government policy. In 2019, after the hydrogen report was published, Birol was in India meeting power minister Raj Kumar Singh when the chat turned to hydrogen. After Birol gave Singh a brief overview, he called Gul from the airport in New Delhi. Within days, Gul had cancelled his vacation, flew to India, and spent an hour with Singh talking about hydrogen.

That kind of access, where you can speak directly and at length to a large economy's energy minister about the potential of a new technology, is invaluable. It's not always possible to say one thing led to another and thus a policy was created or a subsidy was announced, but Motherway has seen time and again how conversations that IEA has with ministries across the world can months or years later result in solid outcomes such as regulations or subsidies.

The IEA also brings together countries that can learn from each other and find mutually beneficial wins. The Danish government, for example, has funded a programme at the IEA to promote energy efficiency in developing countries. That's led to bilateral partnerships with India, China, Indonesia, Mexico and South Africa.

The funding is explicitly to support the goals the Danish ministry for climate, energy and utilities has as part of its plan to support sustainable economic growth in developing countries. But Denmark also happens to be home to some of the world's leading energy efficiency solutions providers – Velux (windows), Danfoss (heating and cooling) and Rockwool (insulation), to name but a few companies. These bilateral partnerships can mean Danish industry benefits from access to a new market while developing countries can find ways to cut energy use, which for most of them means reducing their import bills too.

The IEA's focus on energy efficiency is broader than the programme the Danes have funded. Its mandate for energy conservation goes back to its founding days. That's because energy efficiency is like finding a new source of fuel by, in effect, not using as much as you used to for the same activity. The most recent success for the IEA, says Motherway, is efficient cooling.

Air conditioning is one of the most energy-intensive activities. And not surprisingly, as the world gets warmer and richer so demand for air conditioners rapidly rises. As a base case, the electricity demand for cooling is set to reach three times current demand by 2050.[10] 'It could really wreak havoc,' says Motherway. Investments to support increased demand are likely to increase carbon emissions, because the greatest interest is coming from China and India, countries that have coal-heavy electric grids.

But the IEA's work also found that if more efficient air conditioners are used then it could halve the projected energy use. 'We showed that more efficient air conditioners were available,' says Motherway. 'It wasn't a technology question. It was a policy question.' The results are clear. As a result of the IEA's work both India and China developed national cooling action plans. The IEA is now working with countries in South East Asia to take the lessons the two giants are learning and apply them in the region.

Birol essentially has a gang of analysts – many with PhDs – who can make house calls to any country and help it solve its energy problems. He has been dedicating more and more of them to work on clean energy issues full time.

'I personally feel like we're at a real sweet spot at the IEA,' says David Turk, who was Birol's deputy at the agency until 2021, when he took on the number two role at the US Department of Energy. 'There are very important negotiations that are being done at the UN, but we have the advantage of being substance-driven.' Or, put in simpler words, Turk says the IEA gets stuff done while the UN struggles.

Even the biggest defenders of the UN climate process agree that its negotiations can become unwieldy. Consider what happened at COP26 in Glasgow, the fifth annual meeting since Paris.

All parties to the Paris Agreement had decided to set out more ambitious emissions-cutting goals every five years. In the lead up to Glasgow, scheduled for the first week of November 2021 after the pandemic forced a year-long delay, countries were making the right sorts of noises. By the start of the summit, all ten of the world's largest economies, for example, had announced a net-zero goal after the last hold-out, India, also came on board.[11]

As the two weeks of negotiations came to a close, it seemed that the countries would sign a new pact that could significantly narrow the gap between the Paris target of keeping warming below 2 °C and the track the world found itself on warming beyond 3 °C. The final draft text of the Glasgow pact set out that countries would phase out fossil fuel subsidies and coal use. Then, at the very last moment, things broke down. Before COP President Alok Sharma could bring the gavel down and close the meeting, China and India raised objections.

Both countries wanted to change 'phase out' to 'phase down' – a position also backed by the US.[12] No one could understand why it was so important that the process of adopting text stretching to over ten pages had to be stopped for what seemed a minor, rhetorical change.[13] Despite that, because it came from the world's three largest emitters (who also ranked numbers one, two and six among the largest economies), a new version was drafted to accommodate the request, and put to another vote. Many small island countries protested at how big countries tended to get their way, but ultimately agreed to the change to ensure the rest of the Glasgow pact was intact. Sharma, a UK government minister, was in tears as he talked about how the last-minute interventions went against the grain of the multilateral process.[14]

This kind of seemingly silly stuff often happens at COP meetings. That's because the meetings are run on rules agreed upon in 1992 when member nations adopted the UN Framework Convention on Climate Change (UNFCCC) – which does not include an agreed upon set of rules for voting.[15] That might seem odd, but the outcome is that the only way to get to an agreement at a COP meeting is through consensus where every country must be on the same side (or, at least, not raise an objection), rather than a majority or a supermajority choosing the winning side.

The reason COP meetings end up in this bind is because Saudi Arabia has been blocking the adoption of voting rules since the UNFCCC first proposed them. It's much easier for the fossil fuel giant to slow down the process through requiring a consensus than agreeing to voting rules that would speed up adoption of climate goals.*

That is not to say COP meetings are futile. Decisions that are made by all countries have a tendency to stick. But having to get full consensus typically makes the process slow and the outcome less ambitious. The IEA cannot replace what COP does, but it can help make the process run faster through the work it does on data and policy. Its membership, though small, consists of the most powerful countries. If it can get those countries to agree first, it overcomes one of the biggest barriers for a consensus outcome at COP meetings.

Under Birol, the IEA has transformed. Plenty of credit for those changes should go to the critics of the IEA. The biggest charge has been levelled at the IEA's main product, the *World Energy Outlook*. The annual report uses the agency's rich database and mixes it with sophisticated modelling to project what the global energy system is expected to look like in the decades to come. Given that energy infrastructure can last for decades, the pronouncements the outlook makes can determine the direction of billions of dollars – both private and public.

Even though many other organizations have since developed the capacity for doing the kind of modelling that underpins the *WEO*, none produces an identical report, nor has the sheen the IEA gets as an independent intergovernmental organization.[16]

---

* Technically, the UNFCCC has not defined 'consensus' either. It is down to the president of the convention to decide whether to pass a decision despite some opposition, as has been done in the past, including when the UNFCCC itself was created, in 1992.

'Because it's a multilateral institution, it comes with a certain level of perceived objectivity and professionalism,' says Natasha Landell-Mills, head of stewardship at investment management firm Sarasin & Partners. It's hard to find a high-level government report about energy that does not cite the *WEO* or some other IEA data.

Renewable energy experts watched dumbfounded for several years in the 2010s as solar power prices entered freefall, and yet the *WEO* annually failed to make projections that matched the actual speed of renewable adoption.[17] It had similarly missed the US fracking boom a decade earlier. Part of the issue has to do with the nature of the exercise. The IEA and other modellers emphasize they're not really trying to predict the future but are analyzing what happens if certain conditions continue.

But the pace at which green technologies have fallen in price has surprised even the most optimistic analysts. The poor calls the IEA has made on those technologies have made it a butt of jokes, especially during the time of the annual release of a new *WEO*. Analysts often make charts of previous *WEO* predictions about solar deployment or some such to illustrate just how wrong the IEA has been in the past.

Still, the IEA has been slow in adapting. It wasn't until 2009, about two decades after it was clear that increased fossil fuel use would lead to disastrous climate change, that the IEA produced a model for transitioning away from dirty sources of energy. It wasn't until 2015 that the IEA produced scenarios that would keep warming well below 2 °C, which was the less ambitious goal that was signed off later that year at the Paris COP meeting. And it wasn't until 2021, only months before the COP26 meeting in Glasgow, that the IEA produced a fully-fledged scenario for the world to reach net-zero emissions. That was crucial if warming were to be kept below 1.5 °C. Global average temperatures then stood a touch above 1.1 °C relative to pre-industrial times.

The IEA's reticence to change as quickly as the world demands is because Birol's bosses aren't investors or climate groups, but ministers of rich countries, with much to lose in a rapid energy transition. No one with thirty-one bosses of that ilk – not even a skilled diplomat like Birol – could have repositioned the IEA for climate change overnight.

Unlike other analysts at consultancies or think tanks, Birol has to walk a geopolitical tightrope. He has powerful masters and influential audiences. The US, for example, continues to pay the largest share of the IEA's core budget. Other members may pay millions of euros for specific projects, all money that's crucial to keeping the IEA going and growing. That's why knowing what to say to whom, and when, is a crucial part of the job.

But as the world of energy is being changed, partly driven by climate needs, it's not easy for Birol to control the message. The pronouncements he makes can become headlines in the energy press. If he gets them wrong, those statements also come back to haunt him.

That was especially true given the challenges presented by the presidency of Donald Trump. Under Trump, the US – the IEA's biggest funder – took an actively hostile stance towards climate change, pulling out of the Paris Agreement and overturning many of the emissions-cutting policies that previous administrations had put in place. During the Trump years many IEA members committed to reaching net-zero by 2050 – but not the US. Birol and his team had to wait it out for 2021, when Joe Biden took office and soon after made 2050 the target for his country to reach net-zero emissions. It was only after that that the IEA launched its first fully-fledged look at what such a future would mean for the energy world.

The results were worth waiting for. The IEA's modelling showed that, if the world is serious about keeping warming below 1.5 °C, we need to stop the development of new infrastructure to extract fossil fuels. That means no new coal, oil and gas fields from now on. It is just the kind of line climate activists need to arm themselves with – utilizing the backing of an organization that was once an ally of fossil fuels.[18] Every time an oil and gas company or a government announces a new project, such as when Biden approved the Willow oil project in Alaska, green groups are able to criticize the decision with the full might of not just climate scientists but also the energy world's top analytical group.[19]

The IEA has a crucial role to play in the clean energy transition. With its expanded membership that includes China and India (even if not all are full members), its claim on being a truly international intergovernmental organization is starting to ring true. That said, it remains lean enough to provide a complement to the bogged-down UN bodies. The IEA's mandate to help maintain energy security in the face of volatile fossil fuels, climate impacts and an uneven energy transition is crucial to making climate capitalism work.

For an organization with only a few hundred people and a core annual budget of only €20 million or so (about $22 million), the IEA punches above its weight in the influence it wields in the energy industry and policy arena. As Birol positions it to take on a challenge much bigger than what the agency was first created for, it comes as no surprise to him that the IEA is rubbing some people up the wrong way.

Birol explains it this way. There's a growing gap between the people who care about climate issues and those who care about energy issues. That polarization can be seen in the movement against nuclear power, for example. What were once debates among environmentalists, with organizations like Greenpeace in opposition and small groups like the Breakthrough Institute in favour, have become full-on culture wars. Another extreme is the use of carbon capture technology: this is supported by organizations like the IEA while others like 350.org argue the only way forward is to end the use of fossil fuels.

That's why Birol believes bridging that gap is key to ensuring the transition to clean energy doesn't slow down. It is one of his top priorities for his second term as head of the IEA, which began in 2019 and will last until 2023. A third term is always a possibility, based on performance.

Few positions of international diplomacy and big influence are filled with technically minded people. In that sense, former US Energy Secretary Moniz, who taught physics at the Massachusetts Institute of Technology before he took the job, is probably Birol's closest peer. And Moniz thinks the most value the IEA can add isn't in coming up with more energy–climate scenarios, but instead providing the tools that major economies are going to need for what's likely to be one of the biggest disruptions they will face.

Just as start-ups benefit from accelerators, where they meet mentors who have started companies in the past and investors who are hungry to support new talent, so the IEA performs a coaching role for countries, helping them learn lessons from peers that have successfully implemented climate policies. That's crucial, because addressing the climate challenge will require every country to simultaneously start cleaning up its energy system and finish the job within decades – something that has never been attempted by any country alone or even a small group of countries working together.

In other words, as the world embraces more ambitious climate goals, the IEA becomes more important, not less. Those goals will only be achieved by steeper cuts in emissions while continuing to provide energy, which will require countries to make more complicated decisions. That, in turn, will increase their reliance on advice from bodies like the IEA. When I asked Birol what he wanted his legacy to be, he said that he wanted to build an 'authority that can address the two biggest challenges of the century: climate change and access to energy for all'.

One reason Birol's job has been made easier is because of the unstoppable development of new green technologies. Even as government funding plays a crucial role in helping fundamental research, it is private capital that helps them scale. That's where Birol, who typically deals with government bureaucrats, is getting support from one of the richest individuals in the world: Bill Gates.

6

The Billionaire

When one of the biggest moments in climate history happened in Paris in 2015, Bill Gates was in the middle of it. The annual climate meeting was set to produce an agreement among every country in the world that it's time to cut emissions and stabilize global temperatures. But Gates, a billionaire philanthropist and a private citizen, was there asking those world leaders to put money behind their new-found interest.

Gates, one of the world's richest men, made the case that if countries are serious about tackling climate change then they need to double energy research spending. In return, however, some of the world leaders asked whether Gates would be willing to put his own wealth and ideally those of other billionaires behind commercializing the new technologies that government money might spur on. Without consulting his wealth manager, he agreed to it.

When he returned to his home in Seattle, criticisms of his idea began almost immediately. 'That's a terrible fucking idea,' said Rodi Guidero, who at the time managed Gates's personal investments in start-ups. It's one thing if Gates wants to risk his money – he had agreed to give all of it away within his lifetime anyway. But risking his co-billionaires' cash on nascent clean energy technologies could end up in huge losses and cause Gates embarrassment, said Guidero.

'Why do you think I care about that?' Gates replied. 'If it wasn't challenging, we wouldn't need to do this.' That was the moment of creation for Breakthrough Energy Ventures (BEV), which is now a multibillion-dollar fund that has invested in more than 100 climate start-ups. It was also the start of Gates's unabashed global campaign to get governments to spend on clean technologies.

Since then, despite a climate change denier's stint in the White House and a pandemic, Gates is more optimistic than he has ever been. 'Innovation is going more quickly than I thought,' he said. 'That's why I'm confident that we will solve this thing.'

With Breakthrough Energy, Gates created a framework for how private capital can bring to bear its power to tackle climate change and make a nifty return doing it. The story of how it was built is based on multiple conversations with Gates over the past few years and interviews with the people closest to executing on his vision.

In January 2021, with the pandemic still raging across the world, I spoke to the co-founder of Microsoft from my home in London. Throughout the conversation, Gates spoke while rocking in his chair. The rocking, I later learned, is something he does quite often. There was the occasional pause when he listened to my questions, but the rocking started again when he spoke and it became speedier the more animated he became.

Having made his wealth developing and selling software, Gates has turned his nerdy mind to improving human health, reducing poverty and tackling climate change. When he chooses a problem to solve, he devotes not just billions of dollars to it but a huge amount of time. He gets into the weeds while studying it, which gives him a good command of the subject and strong opinions about what's missing. He's a voracious reader and carries around a suitcase full of books as he flies around in his private jet. He connects ideas across vast subject areas with ease.

'Even getting 50% reduction in emissions by 2050 will be super hard to do, because we're currently on track to increase emissions instead. We'll end up with a whole generation of people who are very cynical,' he says. 'But if we manage to achieve net-zero emissions, it will make the Covid-19 vaccine effort look like nothing. For this generation, it will be the equivalent of ending World War II or stopping Nazism.'

Gates speaks passionately about what he sees as major challenges and solutions to climate change. There were two things that stuck out: his conviction that innovation has a huge role to play in tackling the emissions problem, and his frustration that few understand just how big a role innovation ought to have.

In February 2021 he published a book titled *How to Avoid a Climate Disaster* in which he distilled what he had learned over the previous decade as an investor in and advocate for climate solutions – much of it coming from his work building Breakthrough Energy, a growing organization with tentacles addressing all those problems. The big point: cutting emissions to zero will be hard, but it's not impossible. The case he made was for deployment of a combination of technological innovation, better government policies and greater private capital to make green alternatives cheaper.

But it starts with innovation, and he is candid about his bias. 'When somebody says "Hey, this is a problem . . ." My answer: innovation,' says Gates. 'And I'll say that even if I don't know the specifics.'

In the case of climate change, however, Gates knows the specifics. He often uses the example of clean steel as a way of making the case for innovation. Steel production is the cause of some 8% of global carbon dioxide emissions, and getting to zero requires every source to get there or as close to zero as possible, which means the world will need ways to make steel without releasing planet-warming gases.[1] And that means the world needs to invent a way to make clean steel, then make it cheap enough to compete against the dirty kind.[2]

'Realistically, getting to zero doesn't mean we are going to stop doing the things we are doing – flying, driving, making cement and steel, or raising livestock,' says Gates. Those are what scientists call sectors with 'hard to abate' emissions. Clean alternatives will need a huge amount of innovation along with a significantly growing market share for existing green alternatives such as batteries. Developing and scaling technologies is what Gates wants Breakthrough Energy to focus on.

Since 2015 the organization has grown rapidly.[3] It started with a $1 billion venture fund that got some of the wealthiest people in the world to invest in technologies, each with the potential to cut as much as 500 million tons of CO_2 annually – about 1% of global emissions. Breakthrough's non-profit activities include lobbying governments for policies and funding, issuing science reports to shape public conversations, and creating a platform for private companies to increase demand for clean alternatives.

Almost all of Breakthrough Energy's efforts are in their early days. For example, few of the 100 or so companies that the venture arm has backed has listed publicly – an event that is usually seen as a successful exit for a venture capital investor. Overall, however, there isn't any meaningful economic signal or measurable emissions reductions yet, to judge if the organization is succeeding. But many experts say that what Gates has created is a unique initiative that focuses not just on funding start-ups but also helping to create the ecosystem needed to put their technology to use at an impactful scale. I've been following Breakthrough since the beginning, but the team had done most of its work without opening up about its inner workings. Until now.

The venture arm is pursuing very early-stage technologies, and this makes their investments high-risk. Gates and many of his billionaire friends may not mind losing a few billion dollars on a laudable goal. This dynamic means it might be hard to replicate what Gates has done with Breakthrough Energy. Nonetheless, there are elements of its systematic approach to investing private capital that others can learn from and deploy elsewhere.

Global venture capital spend is rapidly growing and now stands at more than $70 billion annually, exceeding the $30 billion that governments around the world spend on energy innovation each year.[4] Both government funding and venture funds often work to seed many times more private investment as new technologies are scaled up. That's crucial as a contribution towards the trillions of dollars that will be needed each year to build emissions-free alternatives for the entire physical economy – from cement and steel to electric cars and zero-carbon ships.

The success of the Bill and Melinda Gates Foundation served as an inspiration for Breakthrough's creation. From 2000 to 2022, the foundation has spent more than $70 billion on various programmes, the vast majority devoted to global health.[5] It is the world's largest philanthropic organization based on the amount of money it gives out each year, but its real success is in how it has convinced governments and other entities to spend on the causes it finds important.[6]

Consider the success of Gavi, a public–private partnership that helps vaccinate people in poor countries. Gavi says it has provided more than 800 million vaccines since its inception in 2000, saving what it estimates to be some 16 million lives.[7] Up until 2020 the Gates Foundation alone had given about $4 billion to Gavi.[8] The UK, the US and Norway have contributed about $9 billion in total.

The story of the Global Fund to Fight AIDS, Tuberculosis, and Malaria is more remarkable. After the foundation convened partners and provided seed funding in 2002, money from world governments flooded in. As of 2019 the Gates Foundation had given about $3 billion to the project, just 6% of its $49 billion in total funding. About $18 billion came from the US government alone.

Bill Gates learned that it's possible to multiply the impact of the money he spends, if he choses the right cause, builds a strong case for it and seeds the creation of an organization that can spend it right. Those lessons informed his thinking going into the 2015 Paris climate meeting. Almost a year before the meeting, he began conversations with some of his staff about doing more on climate issues.

'It's not like he was new to the climate thing,' recalls Jonah Goldman, who was Breakthrough Energy's managing director until 2022 and who had previously worked for the Gates Foundation and Gates Ventures, the organization managing Gates's personal office. 'People didn't know him as a climate person, but he was putting hundreds of millions of dollars to work to try to solve the problem.' Indeed, by the mid 2010s Gates had already committed money to Silicon Valley firms like Khosla Ventures and Kleiner Perkins, which were behind many of the investments made in clean technology start-ups in the early 2000s and had a proven track record of producing great returns. He had also made direct investments in the nuclear power start-up TerraPower, the battery start-up Ambri and the direct air capture start-up Carbon Engineering.[9]

By 2014, with the Paris meeting on the horizon, support for solar and wind power had begun to rapidly grow around the world, helping to lower the cost of renewables. But the electricity sector accounted for only a quarter of the world's emissions, and it was clear to those in the know – including Gates – that new technologies were needed to address emissions from other dirty sectors. They would come from investments in risky ideas and more funding for research and development.

'Why is nobody focusing on R & D?' Gates asked Goldman in 2014. 'How can we make it a bigger part of the conversation?'

There was a reason not many were focusing on green R & D. Between 2006 and 2011 attempts at funding new climate technologies, such as electric vehicles and biofuels, had proven disappointing.

When the bust came, venture capitalists lost more than half of the $25 billion invested in clean tech over that time frame.[10] Gates's team did a post-mortem and found the problem wasn't that the technologies could not be scaled but that Silicon Valley's internet start-up funding model wasn't right for this new space.

Unlike the technologies used by companies like Facebook and Twitter, which could be rapidly brought to market and then improved upon, those developed by climate start-ups needed much more time to become mature enough for commercial use. They needed more government funding and patient investors, but after the clean tech bust of 2011, private capital was shying away and government funding hadn't stepped up to the scale of the challenge.

Things began to take shape before the Paris meeting. In June 2015 in London, an editor at the *Financial Times* asked Gates about the lack of research on clean energy solutions. The exchange bugged him.[11] He was worried that people organizing the COP meeting were focused on the wrong things. That's when he decided to build his own organization to push the idea forward and decided to call it Breakthrough Energy. The steps that followed show what the world's richest and most famous can do when they are obsessed with solving a problem.

The day after the London meetings, Gates flew to Paris and pitched his idea of a climate tech fund and how governments could support climate tech innovation to French President François Hollande, who was readying his country to host what would become a historical climate meeting. By September Gates was back in Paris meeting with Hollande and Indian prime minister Narendra Modi. Gates wanted to a create an initiative that would start with the biggest countries committing to doubling their spending on energy innovation, an investment he would augment by assembling a group of wealthy individuals and industrialists to make similar commitments to scaling the technologies that government research produces.

Modi proposed 'Mission Innovation' as the name for the government side of the initiative, which would include some of the world's largest emitters; 'Breakthrough Energy Coalition', the private side, would be led by Gates and include twenty-eight investor targets, who came from ten different countries. That included like-minded billionaire tech founders like Mark Zuckerberg of Facebook, Jeff Bezos of Amazon, Reid Hoffman of LinkedIn, Marc Benioff of Salesforce and Jack Ma of Alibaba; industrialists like Mukesh Ambani of Reliance, Richard Branson of the Virgin Group; and legendary investors like Softbank founder Masayoshi Son, Vinod Khosla of Khosla Ventures, John Doerr of Kleiner Perkins and hedge fund giant Chris Hohn.

Both initiatives launched in Paris on 30 November 2015 – the first day of the COP21 meeting. On the day, Gates stood, wearing a too short blue tie and satisfied grin, between Canadian prime minister Justin Trudeau to his right, and US president Barack Obama to his left, with heads of state or high-level representatives from China, India, France, Japan, Indonesia and other major economies flanked alongside. He was the sole representative of the billionaire group, among politicians who lead the countries that have been, to date, responsible for most of the world's greenhouse gases.

The billionaires who didn't make it to the stage had been warned, however. 'Bill told those signing up to the coalition that if you're signing up then you're doing it for really investing,' recalls Goldman. Even if, says Goldman, at the time 'we had no idea how we were going to do it'.

On his return to Seattle, and after his team had seen that Gates was determined to go ahead, Guidero was made executive director of Breakthrough Energy Ventures (BEV) (and, after Goldman's departure in 2022, executive director of Breakthrough Energy). He immediately began working on the structure of the investment vehicle. He found in John Arnold someone who could work with him on the details. Arnold made his name and some wealth as an oil and gas trader in Houston working for Enron during the early 2000s. After Enron collapsed in a financial fraud scandal, Arnold escaped without blame and created a hedge fund where he found even more success, becoming the US's youngest billionaire.[12] In 2012 he retired at the age of thirty-eight and began to work full time on philanthropy.[13] In 2017, Guidero said, Arnold went beyond just committing money to BEV and began advising on strategy. When Khosla and Doerr – two billionaire Silicon Valley investors legendary for their early backing of companies like Google, Amazon and Sun Microsystems – also signed up, the venture fund's priorities began to take shape.

Venture capital funds raise money from investors called limited partners: they are 'limited' because they are typically not involved in the day-to-day management of the fund, and their liability is capped at the sum they contributed to the fund. The management of the fund is done by a team that has raised the money for a specific thesis, say, in BEV's case technologies that will tackle climate change. The 'venture' part means the money is invested in start-ups at an early stage, with an eye on growth potential. The risk is high, but so is the possibility of reward. Most start-ups in the fund's portfolio will fail, but some are expected to make a spectacular return that can more than make up for the losses elsewhere. It's not unusual for venture funds to return tens of times the money invested on successful bets and even hundreds of times on wildly successful ones. So far, however, the biggest successes among venture funds have been for those investing in internet start-ups. Climate tech companies need more money to scale and are a much riskier investment.

Typical venture funds also have a ten-year life, which means that at the end of the ten years limited partners get to cash out. That's only possible if the portfolio companies have either publicly listed, been acquired by an investor, merged with another company, or if another investor has bought out the venture fund's shares in the company.

The fund BEV put together would be different, lasting twenty years.[14] After speaking with Arnold and others, Guidero understood that a ten-year lifespan just wouldn't work in this context, because most climate start-ups work on ideas rooted in hard science and technology, which typically take more than ten years just to prove commercial viability and many more years after that to start selling their wares. BEV's investors would have to be patient and understand that their capital was going to fund high-risk, high-reward, long-term climate solutions.

By 2017 BEV had raised about $1 billion, making it one of the world's largest climate-focused venture funds. News articles about the fund began to appear, followed by pitches from thousands of start-ups. Even though $1 billion was a large sum, the team was quickly overwhelmed with start-ups that seemed worth investing in.

Most VC funds start with a clear thesis about what companies they will look to invest in and then raise money from those who find the thesis compelling. In BEV's case, Bill Gates's name and address book was enough to raise the money. The next crucial step for BEV was to create a framework that would help narrow down the fund's remit. A team of science experts was assembled in 2017, and it set a clear goal: the fund would only invest in start-ups with the potential to reduce greenhouse gas emissions by at least 500 million tons each year (approximately 1% of global emissions at the time) when fully scaled up.

BEV had in-house expertise to assess potential investments: many of its members had PhDs or extensive experience working in specific technology areas. But BEV would also use external experts to evaluate the people behind the start-up, scaling-up plans and the team's ability to execute on those plans.

Combined, the size of its fund, the technical chops of its team and the high bar it set for emissions-cutting technology made the

BEV fund unique in the clean tech space. But it still left me with a lingering question: if losing money was okay for Gates and his billionaire friends, why launch a venture fund that aims to provide returns that are much greater than the sum invested? 'The scale is so large that you have to have a much broader group of people investing,' says Guidero. As wealthy as the billionaire class might be, they cannot meet the challenges of the climate crisis on their own. Reaching net-zero emissions by 2050 will need an invest-ment of $4 trillion in the energy system annually between now and then, according to the International Energy Agency.[15]

Structured as a venture fund, Breakthrough Energy could do for private capital what the Gates Foundation achieved with government funding: multiply the initial sum many times over by convincing others why the ideas that Gates's team chooses are worth backing. BEV was designed to make money, of course. But it had a higher purpose. BEV's seed funding of early-stage technologies would help lower the risk for other types of inves-tors to back them as well. What Guidero was saying is that the task of cutting emissions to zero means changing much of the world's physical economy with technologies that don't produce emissions, and philanthropy or government funding alone are not enough for such a big change. Gates needed to tap the biggest lever in capitalism: private capital itself.

Eric Toone says it took him one second to decide whether to accept the job at Breakthrough Energy Ventures. In 2016, when the call came, Toone was a tenured professor of chemistry at Duke University and leading the university's innovation and entrepreneurship initiative. What mostly interested the BEV team, though, was his previous work at the US Department of Energy.

Following the 2008 financial crisis, then-president Barack Obama created an $800 billion programme to help pull the US economy out of recession. A small fraction of that money, about

$400 million, was allocated to the creation of the Advanced Research Projects Agency – Energy (ARPA-E). The agency, housed within the Department of Energy, was modelled after a successful iteration in the US Department of Defense called DARPA, which is credited with having created the networking framework in 1969 that laid down the foundations for the internet.[16] ARPA-E's main job was to fund high-risk, high-reward energy innovation in government labs, universities and private companies. Toone was made its first deputy director of technology.

'Addressing carbon is almost an existential issue for humankind, and I don't think I had that beaten into my head until I was at ARPA-E,' says Toone. 'Working at ARPA-E changed everything about my life.'

But the government job came with restrictions. ARPA-E funded interesting ideas in their early stages, but at later stages – when far more money was needed to scale them up – funds often ran out, and private capital was needed to decide which of those ideas would be worth continuing to finance.

That's why Toone jumped at the offer to join the BEV team. 'Government is a pretty blunt tool. Picking winners and losers is not the role of the government,' says Toone. 'At BEV, I saw an opportunity to continue what we started at ARPA-E but use a scalpel instead. It's explicitly my job now to pick winners and losers.'

Toone wasn't the only one who wanted to continue the work started at ARPA-E. In all, BEV has a staff of some thirty.* That includes five former employees of ARPA-E, who all have a science PhD. Toone is BEV's technical lead, which means it's his job to get BEV to invest in companies with technologies that are

---

* Typical VC firms are much smaller, with many having fewer than ten fully employed staff.

not just scientifically sound but also likely to succeed. He's also responsible for ensuring that BEV's bets are made across five key economic sectors: electricity, transport, buildings, manufacturing and agriculture. Each sector has a staff lead, but 'at the end of the day, the buck stops with me,' Toone says.

Toone's early tasks were to deal with the huge number of pitches from start-ups that came in following the media's coverage of BEV's $1 billion fundraise. 'I'm sure we looked at 4,000 companies by the time we'd done our first twenty investments,' says Toone. That list of twenty was completed in 2019, and, as promised, it covered technologies in many different sectors.[17] In electricity, there were six start-ups in energy storage, two in solar, one in geothermal and one in nuclear fusion. Transport got one investment in an electric vehicle start-up. Buildings had one start-up each doing rooftop solar, smart electric metres and water usage. Manufacturing got one start-up each working on cement and steel. Agriculture saw four start-ups in sustainable food production.

Electricity-related start-ups garnered a disproportionate interest in the first lot. Electricity is about a quarter of global emissions. But that made sense at the start because BEV was being reactive to the pitches. Since the Paris Agreement, the number of climate start-ups launching has ballooned and so has the money supporting them. Following the clean tech 1.0 boom, annual investments in climate start-ups stood at about $400 million in 2013. That number rose to $16 billion in 2019 and $50 billion in 2022.[18]

But sectors like manufacturing lag far behind in the innovation landscape. That's partly because the companies in those sectors, such as steel giant Arcelor Mittal or behemoth Heidelberg Cement, operate in a business that has small profit margins. Crucially, any change to underlying technologies is resisted with excuses ranging from conservativism in thinking to unthinkable expenses. 'We knew in the back of our mind that we're not going to be successful if we don't tackle big industrial areas like cement, steel and cattle,' says Toone. 'Those are the unsexy parts.'

With the first tranche of investments done, BEV was ready to move from being reactive to pitches to being proactive to sniffing out start-ups in priority areas. Cement is a particularly intractable problem, and BEV's approach here is a good illustration of how it tackles sectors with hard-to-abate emissions.

Making cement causes as much as 8% of global emissions, and there aren't yet any economical technologies capable of significantly reducing the product's carbon footprint.[19] Cement plants can last for fifty years or more, lowering the turnover rate for new technologies to enter the market.[20] Cement is also a cheap and bulky product, which means plants have to be widely distributed geographically. One industry expert said that a cement plant is only able to meet the needs of customers within a 300-kilometre radius. That's because the cost of transporting a bulky product longer distances quickly becomes unviable. Thin profit margins have also forced consolidation within the industry, with a few companies controlling the vast majority of global production. That's because large companies are able to distribute the risks to their business and thus tap cheaper pools of capital.

The cement industry also enjoys laxer regulations on emissions compared with, say, power plants. In Europe, for example, offshore wind power is getting cheaper, which allows governments to apply more pressure on coal power plants to cut emissions and eventually drive them out of business. There aren't yet cleaner alternatives to cement, which remains vital to the economic growth of a country, and thus the industry doesn't feel under pressure to innovate.

'Cement has been around since antiquity, but it hasn't been the subject of significant scientific study,' says Toone. 'It's hard to think of a topic less sexy than cement.'

Cement as a material is also quite hard to study scientifically. Any two batches of cement are not chemically identical. It's also amorphous, which means its constituent atoms are not well structured, as you would find in a crystalline material such as the metals used in lithium-ion batteries, that you can probe with methods such as X-ray crystallography. Crucially, cement has been doing its job so well that scientists have never been called on to fix it – that is, until it became clear that the world needed to reach net-zero emissions.

So, when BEV set out to make cement greener, it started by analyzing which aspects of the cement-making process produce emissions. First, limestone is mined and transported to the cement plant, where it is crushed, mixed with clay and fed into a kiln. The kiln is typically a long, horizontal, rotating tube that burns coal to temperatures as high as 1,400 °C. That process first converts limestone ($CaCO_3$) to lime (CaO), which releases lots of CO_2. The lime (CaO) then reacts with silica (SiO_2) in the clay to form calcium silicates – which, in aggregate, is called the clinker, and is the binding element in cement. The clinker is then ground again to become a fine powder and mixed with filler materials, such as limestone and silica, to make cement (which is about 70% clinker).

That is then packed and transported on lorries or ships to the customer. At a construction site, the cement is mixed with gravel and water to make concrete. In the process, some portion of the clinker – that is, the calcium silicate – is converted back to limestone, absorbing carbon dioxide from the air. The newly formed limestone is part of the concrete mix and gives it strength, but cement's carbon footprint remains big because only a small fraction goes through that conversion to reabsorb CO_2.

Except for this last step, all aspects of the process produce CO_2. The vast majority – 90% – of the emissions, however, come from what happens inside the kiln. The chemical step of converting limestone into lime is responsible for about 50% of the emissions of the entire process and about 40% comes from burning coal for heat inside the kiln. The remainder comes from carbon-intensive electricity or fossil fuels used in other steps, such as grinders to make the powder and vehicles to transport the product.

BEV is currently evaluating technologies that would create cleaner kilns. It won't disclose the names of all the companies behind them, because those are commercially sensitive details. But Carmichael Roberts, BEV's business lead, confirmed that one of the top candidates ought to be the Australian start-up Calix.

The conventional kiln burns coal, limestone and clay together, and releases a mixture of nitrogen, carbon dioxide and oxygen. It's possible to build a carbon capture facility that separates the gases and then buries the greenhouse gas underground, but that's quite expensive (see Chapter 7). Instead, Calix's kiln only produces carbon dioxide. By eliminating the need to separate the gases, the cost of capturing the carbon – and then compressing and burying it underground – drops significantly.

The way Calix does it is so simple it is ingenious. Its kilns are made up of two concentric barrels: the outer barrel burns gas or uses electricity for heating, providing heat to the inner barrel that contains the limestone and clay. Thus the only chemical reaction inside the kiln is the conversion of calcium carbonate ($CaCO_3$) into calcium oxide (CaO), which releases a pure stream of carbon dioxide (CO_2). If the heating is done with renewable electricity then cement emissions can be cut by as much as 90%. The start-up is currently retrofitting a kiln in Germany; once up and running, in 2023, it will capture 100,000 tons of carbon dioxide each year, Calix says.[21]

Not all companies will be willing to retrofit kilns or acquire a site near the plant where captured carbon dioxide can be buried. In response, BEV is also investing in companies working on reducing the emissions of other aspects of the cement-making process. Ireland's Ecocem, for example, is working on a way to use only 20% of clinker in its mix, compared with the 70% standard. BEV has, to date, invested more than $25 million in the company.

'We appreciated the depth and rigour of science and engineering behind Ecocem's approach,' says Toone. Ecocem was founded in 2000, when it began selling granulated ground blast furnace slag, which is a long phrase to describe a by-product of the steel industry that has chemical properties similar to clinker. Ecocem realized that instead of dumping the stuff in a landfill, they could sell it for use as the binding element for cement mixes. As the political will to reduce carbon emissions gained momentum in Europe, so companies in the region began to buy the slag powder from Ecocem to make lower-carbon cement.[22]

But scientists have known for a long time that not all the clinker or slag inside cement is used for its binding properties. Cement companies use clinker or slag for other purposes, such as keeping the concrete mix wet for longer or maintaining the mixture's viscosity. With reductions in carbon dioxide gaining momentum in Europe, in 2013 scientists at Ecocem started working on ideas to drastically reduce clinker content in cement.

After seven years of research, Ecocem found the right mixture that was 20% clinker, 30% slag and filler materials for the rest. It wouldn't share what the filler material is made up of, because that's proprietary, except to say that it has a tiny carbon footprint and its job is to maximize the activation of clinker and slag so they are more effective binding agents. BEV's cash and its global connections are now helping Ecocem to apply for licences to sell their eco-friendly cement in Europe. Soon it will build small plants in North America and Asia to showcase its cement to potential customers in those regions.[23] It will also look to license its chemistry to existing cement makers, which would make it possible to scale its technology faster.

Another bet BEV has made is on the US start-up Solidia Technologies.[24] It, too, lowers the clinker content in cement – but in Solidia's case, it replaces it with clay. Clay is fine-grained soil that's typically found enriched with minerals similar to those found in cement. To ensure that the concrete made from Solidia's cement is as good at its job as that made from ordinary cement, the company cures its concrete mix in carbon dioxide. This injection of CO_2 forces more of the clinker to convert to limestone and thus improves the concrete's strength. Instead of the concrete taking twenty-eight days to set, the process is complete in a day and Solidia claims that its clay-based concrete blocks have the same strength as carbon-intensive alternatives. Overall, Solidia claims its cement has a carbon footprint that is over 50% lower than the typical product on the market.

Finally, BEV has invested in Canadian start-up CarbonCure Technologies. The company doesn't do anything to tweak the content of cement or concrete. Instead, it simply uses carbon dioxide for curing cement. That increases the amount of clinker converted back into limestone. Thus, concrete that has gone through CarbonCure's step has an overall decrease in carbon dioxide – though it is a small decrease, less than 10% of its overall emissions.[25]

Between Calix (or a similar start-up), Ecocem, Solidia and CarbonCure, BEV will have bets on many of the steps in the cement to concrete-making process. None are guaranteed to work at scale, but that's why, Goldman says, it's essential to have 'multiple shots on goal'. In fact, BEV's plan is to have that for every major carbon-intensive industry.

While technology development is crucial, scaling a start-up requires that there be enough of a market for the green alternative. In fact, if there is a market for it then even the big incumbents are more willing to spend on innovation. As the carbon price in Europe has crept beyond €80 a ton (about $90), more and more cement companies are looking for cleaner alternatives.[26] The production of a ton of conventional cement releases about half a ton of emission. 'Even a modest carbon price would push the industry to become cleaner,' says Toone. 'That's because cement is $100 or so a ton delivered to my door.'

This reliance on external levers is the biggest risk factor for the BEV portfolio. The green technologies and clean products BEV-funded companies are developing will rely on the demand being there for them when they reach scale. That demand will come from either pricing carbon, enacting other types of regulations to rein in planet-warming gases, pressure from investors for cutting emissions, or companies wanting greener products because customers are demanding them.

Given the number of catastrophes climate change is unleashing, there's probably a good chance that one or many of those levers will create the necessary demand. But Breakthrough Energy isn't simply waiting. It has launched a number of non-profit arms that focus on pushing on those levers.

A small BE team in Washington, DC lobbies the US government to deploy policies that Gates believes are crucial. When I spoke with Gates in January 2021, he pointed to the Department of Energy's $160 million grant towards advanced nuclear reactors as a success story of this effort.[27] It helped that in October 2020 his personal-office advocacy led to his nuclear power start-up TerraPower getting an $80 million grant towards showcasing new nuclear technology.

The biggest success of such lobbying efforts came in 2022 when the US passed the Inflation Reduction Act, a cleverly named piece of legislation that's really a climate bill. Work on a large bill that would cover climate, health and taxes had begun as soon as Joe Biden became president, in early 2021. However, because of the quirks of the US Congress and the slim majority Biden's Democratic Party had in the Senate – the upper house of the legislature – passing the bill would have needed all fifty senators to vote in favour. But Senator Joe Manchin of West Virginia, perhaps the most conservative of all the Democrats, had strong ideas about what should and should not go in the bill. He was also earning millions of dollars a year from a company in his state that supplied coal.

As soon as Biden came to power, environmental groups, lobbyists for green technologies and anyone who cared about a climate bill knew that Manchin would prove to be the deciding vote. Almost immediately everyone involved began lobbying the West Virginia senator to accept that their favourite policy was crucial and why he must vote for a climate bill.

The full story of how the climate bill finally became an act in August 2022 is probably a book-worthy subject, but what's most relevant here is that Gates and the Breakthrough team were among those lobbying. Gates personally met and spoke with Manchin numerous times over the eighteen-month period while the bill was being negotiated.

Manchin, who had been a businessman before turning to politics, was more receptive to business-friendly voices like those of Gates than he was to climate activists who kayaked around his houseboat and blockaded coal power plants in West Virginia. Gates's team made the case that even as he was investing in start-ups, no single person's fortune was enough to help scale those technologies. The only way American innovation could be unleashed was through the US government's purse. Specifically, Gates and others wanted the government to approve tax credits, which would reduce or compensate for the green premium attached to earlier-stage technologies like green hydrogen, sustainable aviation fuels and carbon dioxide removal.

The final bill contained precisely that, namely tax credits totalling nearly $150 billion to be used over the course of the next decade. That's more than twenty times the sum that Gates's Breakthrough Energy Ventures plans to spend on clean technologies. Yet again, Gates had succeeded in bringing in many times more money from the government and from other individuals invested in BEV.

Other Breakthrough teams work on scientific reports. The idea is to get government regulatory bodies to value the solutions that the technical minds at Breakthrough consider potentially important. Its first report, published in February 2021, looked at the potential impact of extending the electrical grid out across the US, which, it argues, would enable the movement of renewable electricity from the middle of the country, where there's ample land for solar and lots of wind on the Plains, to the centres of consumption, on the east and west coasts.[28]

It's a compelling case. Currently, US states plan to spend some $360 billion on clean energy deployment up to 2030. But that would only reduce US emissions by an estimated 6%. Breakthrough's science experts found that spending $200 billion on transmission could help the US halve its emissions while allowing its energy grid to absorb a lot more intermittent power from solar and wind. Breakthrough is also funding young researchers to start companies through the Fellows program.

Then there's Catalyst, a programme that Breakthrough began building in 2021. Its aim is to provide grants to and equity investments in climate start-ups that are building their first large-scale plant. That would help lower the risk of investment in the company, which would make getting loans from private banks cheaper. In 2022, Catalyst made its first grant of $50 million to help fund the first large-scale plant making sustainable aviation fuel. The company behind it is LanzaJet and it claims that the fuel it will make from ethanol derived from plants will have 70% fewer CO_2 emissions than conventional jet fuel.[29]

Similarly, large companies could set up advanced purchasing agreements. Say a green aviation fuel company is looking to build its first plant. If executives at Google committed to purchasing a certain amount of fuel to lower the emissions of their business travel from this company before even the plant is built, then it helps lower the risk of investment in the company. Companies would do this because it would provide certainty for their decadal plans to reach net-zero emissions, which more than 20% of the world's 2,000 largest companies have already committed to.

The upshot? With the backing of a grant and purchase orders from established companies, start-ups will be able to secure lower-interest loans for the hundreds of millions of dollars that will be needed to build large-scale plants. Breakthrough Energy is hoping Catalyst will serve all kinds of green start-ups, including the ones BEV is betting on.

Bill Gates is using Breakthrough Energy to show how 'patient' capital is crucial to scaling new technologies, be that from governments or private individuals. But money, even in very large sums, is not enough. Climate capitalism also requires creating an ecosystem of non-monetary support for innovation to happen and technologies to scale.

It's too early to judge whether Breakthrough Energy has succeeded. There have been some early successes, however. Its 2018 investment in battery start-up QuantumScape, which we encountered in Chapter 3, bore fruit when the company publicly listed in 2020 and, in December that year, reached a market capitalization of $50 billion – more than the valuation of car maker Ford at the time.[30] But most of Breakthrough's companies are quite far from reaching that kind of maturity. The non-profit arm is younger and its successes harder to quantify.

Not everyone is impressed by the bets the company has made so far. 'Breakthrough Energy Ventures came out of the gate with a bang, promising to invest patiently in climate solutions others had not,' says Julio Friedmann, chief scientist at consultancy Carbon Direct. 'We're still waiting for that day.'

Still, from an idea in Bill Gates's head to an organization employing more than 100 people, Breakthrough Energy has become a force to be reckoned with. Carmichael Roberts, BEV's business lead, says that his highest achievement so far has been hearing from investors looking at a start-up investment ask one question repeatedly: 'What does Breakthrough think?'

With more than 100 start-ups in its portfolio, Roberts is pleased that BEV has been able to invest across the five economic sectors.

There are still areas, such as carbon capture, green hydrogen, clean steel, zero-carbon aviation and emissions-free shipping, where BEV has not yet found enough exciting start-ups. Rather than wait for pitches to come in, Roberts and Toone are taking on that challenge by launching companies of their own.

Reflecting on the progress so far, Gates says that he started working on what became Breakthrough because his time at Microsoft and the Gates Foundation convinced him that innovation is key to solving many of the world's important challenges. 'But I was a little reluctant to say that my optimism about innovation in software and medicine could have mapped onto a deadline-drive, total replacement of the physical economy,' he says. Breakthrough Energy is his systematic attempt at proving himself wrong.

Funding lots of start-ups is a proven way to disrupt an industry. Most will fail, but some will grow big enough to challenge incumbents and even displace them. But given the urgency to cut emissions, start-ups may not always get enough time to do that across all sectors of the economy. That's where government policies and investments, such as through the Inflation Reduction Act, can make a huge difference. However, there is much that governments can learn from past successes and failures. Consider the fate of the technology in which Donald Trump was putting his faith: carbon capture and storage.

7

The Wrangler

Climate change is a problem of what is up in the air, but Julio Friedmann's focus is oriented in the opposite direction: under the ground. For that's where the story began, as fossil fuels were dug up and burned. And, he argues with conviction, that's where the story should end, as carbon dioxide is buried back down.

Friedmann is a geologist, a storyteller of the subsurface. Bore down through the Earth's crust, and he can walk you through the history of our planet over hundreds of millions of years. In fact, sometimes there is no need to drill. Just as ultrasound can be used to peer inside a soon-to-be mother's belly, revealing the tiny toes of her growing baby, so geologists can send soundwaves into the ground. Different layers of rock then reflect the sound back differently, revealing what they are made of and the secrets they may hold.

And in the same way an ultrasound can change a parent's life forever, a readout changed Friedmann's life. In 2001, during a routine meeting at the University of Maryland, he saw ultrasound images of a piece of crust off the coast of Norway. State-owned oil company Statoil (now Equinor) was capturing carbon dioxide from its oil and gas extraction, compressing it into a liquid, and then pumping it down deep underground. Often oil and gas fields also contain carbon dioxide, and the company had figured out a way to reinject the greenhouse gas underground, even as fossil fuels were extracted.[1]

'I looked at the picture and – boom – I got the whole thing,' he says. Injected carbon dioxide in a rock changes the way the sound is reflected and shows up, 'like a Christmas tree', on the seismic reflection readout. 'Once the CO_2 was injected underground,' he says, 'it's just going to stay there for 50 million years' – just as oil remained trapped at those depths, until humans poked a sophisticated straw in and slurped it up.

By the time Friedmann grasped the potential of this technology to reduce global-warming gases, the oil industry had already commercialized it for decades albeit for a completely different reason. The first use of carbon capture and storage (CCS), as the technology is called, was not to reduce the amount of greenhouse gases dumped in the atmosphere. Rather, it was to increase the production of oil from aging fields.

The fact that CCS was invented by the very industry whose incentives have been aligned with promoting the use of climate-polluting products, rather than cutting emissions, has plagued the technology's development for climate good. That's despite the fact that the technology can, in theory, be applied to cutting emissions of existing facilities, such as power plants and oil refineries, without hindering how they operate, and the fact that it is likely to be an economically viable option to decarbonizing heavy industry, such as cement and steel, which together make up more than 10% of global emissions.

As we have delayed lowering emissions so CCS has become more important in the effort to avert the worst impacts of global warming. In fact, the delay has been so great that most futures modelled by the IPCC to meet the goals set out under the Paris climate agreement require the use of 'negative emissions' (see Chapter 8).[2] That involves the use of CCS technology to capture carbon dioxide that humans have already emitted directly from the air and then bury it underground.

'We live in a world that is fast and dynamic,' Friedmann says. 'But geological systems are not.' That perhaps can be said of CCS, too. After years of efforts from environmental groups that want to scale the technology, there are now some two-dozen plants capturing and pumping 40 million tons of carbon dioxide into the bowels of the planet. That's a relative pittance, accounting for merely 0.1% of all annual global emissions. The world will need to build infrastructure capable of capturing and storing 100 times as much CO_2 within the next few decades to keep global warming to below 2 °C.

Friedmann, who now works as chief scientist at the consultancy Carbon Direct, understands better than most both the urgency of scaling the technology and the multiple forces holding it back. In a thirty-year career he has worked for the oil giant ExxonMobil, as a researcher at a university and then in a secretive national lab, as a high-level bureaucrat in the US government with a big annual budget, at think tanks shaping policy and in a start-up scaling climate solutions. In all but one of those jobs he's been focused on, as he puts it, 'putting as many tons of carbon back into the ground' as he can. That's why he calls himself a 'carbon wrangler'.

Friedmann has worked all his life in the US, which remains a leader in deploying CCS, even as it has lost the lead in other crucial climate technologies, such as batteries and solar. Crucially, CCS's slow development exposes what's wrong with capitalism as it exists and what needs to be done to reform some aspects of it. If CCS is as necessary as the scientists tell us it is, then Friedmann's experience over the last two decades points to the kinds of lessons that will need to be implemented globally and fast.

Carbon capture was first developed in the 1930s in order to remove a contaminant in natural gas.[3] Many gas fields also contain carbon dioxide (and hydrogen sulphide) – the mixture is called 'sour gas' because CO_2 is slightly acidic in nature and thus sour to taste, like citric acid in lemon juice. Those gases need to be separated before pure natural gas can be sold for commercial use.

One way to neutralize an acid, as students learn in chemistry lessons, is to use a base; the resulting solution will be nothing but salt. To separate out acidic carbon dioxide from any mixture of gases, scientists use a base or alkali called amine (which is essentially ammonia – one nitrogen atom attached to three hydrogen atoms – but with one or more of its hydrogen atoms replaced by something else). As the gas mixture passes through the amine solution, carbon dioxide molecules get caught, like iron filings on the surface of a magnet, forming a salt. Out of one end comes a CO_2-free stream of natural gas ready for use in homes for heating or industrial use; out the other is carbon dioxide, released by heating the amine salt in a separate chamber. The CO_2-free amine is then ready to capture more carbon dioxide from additional gaseous mixture run through the system.

In the early days the expelled carbon dioxide was simply dumped into the atmosphere. Then, in the 1970s oil companies found a use for the greenhouse gas.[4] As oil fields age so their yields decline. That forces engineers to get creative. One method to enhance the recovery of oil is to pump steam underground, increasing the pressure and pushing out more oil. But because oil and water don't mix, steam's ability to draw more oil simply based on increased pressure hits a limit after a while. That's when someone had the bright idea to use carbon dioxide.

In its liquid state, carbon dioxide acts in the same way soap does on food-stained clothes, it dissolves oily matter. It does an even better job than pressurized steam at pushing out oil trapped in tiny pores of sedimentary rock. For every ton of carbon dioxide pumped into these fields, about three-fourths of it comes back up with newly extracted oil. The rest takes the place of the oil it just pushed out, getting trapped in the same pores where oil had been sitting tight for millions of years.

That inadvertent benefit is what scientists want to exploit on a much larger scale. Dotted through the earth's crust are many geological formations that can hold CO_2 in much the same way fields trap oil and gas. In fact, there's enough space to put in trillions of tons of carbon dioxide, or much more than humans have put into the atmosphere burning fossil fuels in the last 200 years.

Oil company Equinor, which is majority owned by the Norwegian government, pumps carbon dioxide into underground brine fields found offshore.[5] As liquid carbon dioxide fills up the pores, it pushes out harmless salt water. In Iceland, public utility Reykjavik Energy captures carbon dioxide from a geothermal power plant and then buries it into basaltic rock. At a depth of about 700 metres, the greenhouse gas reacts to form minerals – turning CO_2 into rock in less than two years.[6]

But neither CCS's fifty-year history nor its climate potential are widely known or understood, even among many top decision-makers. 'There are plenty of people on the left and the right, who will tell you that this technology is not ready for prime time,' says Friedmann. 'That's just hogwash.'

Some sceptics fear the unintended consequences of leaks, such as unmonitored escapes of large amounts of greenhouse gas. Others compare CCS to fracking for natural gas, which involves injecting liquids at shallow depths and has a bad reputation among environmentalists for causing contamination of water and small earthquakes. CCS typically involves storing carbon dioxide deep underground, which eliminates fracking-type risks. Still others, who fall for the 'not in my backyard' rhetoric, simply oppose any activity in their immediate surroundings, regardless of its safety or utility. But years of scientific studies suggest these worries are unfounded.

Weeks before countries were set to meet in Paris in 2015 to settle on the now famous climate agreement, a group of forty CCS experts from around the world wrote an open letter to the United Nations. It said: 'As geoscientists and engineers representing decades of scientific research worldwide, we would like to reassure the United Nations . . . that the geological storage of carbon dioxide . . . is safe, secure and effective, and we have considerable evidence to show this.' And, as academics do, they went on to provide a long list of peer-reviewed studies as the evidence.[7] But those facts haven't swayed the opinion of enough people – yet.

The origin story of CCS is likely its biggest challenge in gaining widespread support. Just as nuclear power remains inextricably linked to nuclear weapons and ghastly fears of radioactivity, so CCS hasn't been able to shake off its link to the fossil fuel industry and attempts by these powerful companies to sow doubt about climate science.

Many environmentalists have concluded that any activity of fossil fuel companies purporting to cut emissions or build green technologies must be viewed with extreme scepticism. On Twitter, some have started 'green trolling' these organizations: whenever an oil company tweets from its corporate account, they reply with an article showing the company's historical denial of climate change along with 'this you?' Others would prefer to 'cancel' these companies out of existence altogether and often reply to those tweets with #Abolish[InsertCompanyName].[8]

That's why, 'From day one, when I started working in carbon capture, I consistently said that CCS is not a coal technology and it's not an oil and gas technology,' says Friedmann. 'It's an emissions-reduction technology.'

Though oil and gas companies were the first to master CCS, most of the interest in the technology around the turn of the century was coming from coal companies – for two reasons. First, coal is the dirtiest of fossil fuels. For every unit of electricity generated, burning coal produces nearly twice as much carbon dioxide as natural gas (assuming no leaks in the natural gas supply). That meant the pressure on reducing emissions was the greatest on coal miners and utilities with coal power plants. Second, coal was the top fuel of choice for meeting increasing electricity demand. That's because natural gas was expensive at the time, with the shale revolution yet to happen. Coal power plants generated 40% of the world's electricity, and its burden on global emissions was growing rapidly.[9] Solar and wind power, which had to be heavily subsidized, barely registered in the broader picture of global electricity production then.

All that made coal companies enemy number one among climate activists. But when they went, cap in hand, to the oil and gas companies for help making CCS work, they were disappointed: many oil and gas companies had come round to accepting climate science and did not want to be associated with coal companies that still had a notoriety for denying climate change. 'It's funny to say this now, but back then, oil and gas companies were worried about reputational risk,' says Friedmann.

On the other hand, they did see an opportunity: the coal industry, with its awful carbon footprint, had no choice but to spend money on technologies like CCS in an effort to clean up its act. That meant oil companies could, in theory, sit back and watch coal companies invest in CCS, wait for the technology to travel up the learning curve and fall down the cost curve. After that, they would be in a prime position to benefit once the companies also get regulated to deploy CCS on their own facilities. Eventually, over the next two decades, two coal power plants would successfully build CCS units to reduce emissions. One would receive support from the oil company Shell, but in both cases the carbon dioxide buried was taken up by less well-known small oil companies.

Friedmann watched these debates within the fossil fuel industry with dismay: 'as much as I want to accelerate the clean energy transition, I do not see a world without fossil fuels for the next fifty years,' he says.

One reason is that fossil fuels are very cheap. Even after accounting for oil price swings, it's cheaper to buy a litre of crude oil than it is a litre of Diet Coke. Coal is cheaper still. Oil and coal are easy to transport, and there's vast existing infrastructure available to extract energy from these dense sources. But Friedmann also understood that continued use of fossil fuels without abating emissions would mean blowing past all climate targets and inviting avoidable catastrophes.

In 2013 he got his shot and learned the hard way what's needed to build large-scale CCS plants. The Office of Fossil Energy at the US Department of Energy (DOE) hired him as its principal deputy assistant secretary, a position that came with a $600 million annual budget.[10]

The DOE is a quiet, revolutionary force in shaping science and technology that has global impact. It has helped make batteries for spacecraft, advanced medical research by helping the development of MRI scanners, and discovered eight elements on the periodic table. Even under President Donald Trump, who once called climate change a hoax, the DOE continued to fund the development of new technologies. The department's secretary typically has an annual budget of more than $30 billion, with vast sums doled out in support of cutting-edge technologies from lithium-ion batteries to CCS.[11] It funds top scientists in the US's secretive national labs and gives out grants to private companies to deploy technologies too risky for commercial investors.

It was Friedmann's opportunity to put the experience of his entire career to use. Working at ExxonMobil had helped him see how oil companies think about emissions. Working as a researcher at the University of Maryland in College Park and then at the Lawrence Livermore National Laboratory in California made him an expert at understanding CCS technology. When he was offered the DOE job, he accepted it because he also knew there was no way to scale CCS without government intervention.

Friedmann's main task was to build eight large-scale carbon capture projects, or at least get them initiated. By far the best place in the world to try was the US. After all, the country's home-grown oil industry invented and mastered the use of injecting CO_2 into oil fields to enhance production. In the process the

US had trained people to deploy the technology and invested in infrastructure, such as pipelines to carry carbon dioxide over long distances, setting the table for future projects.

Two of the projects Friedmann worked on – one a success, another a failure – provide a good window on what he learned scaling CCS technology.

Let's start with the failure. In the early 2000s, when natural gas prices were high and renewables very expensive, it seemed almost certain that the use of coal for generating electricity would keep growing. The DOE wanted to fund new types of coal power plants that would have carbon capture built into them from the ground up. That's what led to the Kemper Project.

Among the evergreen forests of Mississippi, the DOE would support the building of a unique power plant that would turn coal to synthetic gas with reduced greenhouse gas impact. The process involves heating coal at very high temperatures in a chamber with oxygen and steam. That breaks coal down to its molecular components while only partially burning it, resulting in a mixture of three gases: carbon monoxide, hydrogen and carbon dioxide. The carbon dioxide would, in theory, be captured and transported to an oil field, where it would be injected into the ground and thereby increase extractive yield. The mixture of carbon monoxide and hydrogen, called synthetic gas, would be burned, like natural gas, in a power plant to produce electricity.

After years of planning and revisions of cost estimates, in 2010 Southern Company, a multi-state gas and electric utility, announced that the plant would be operational by 2014 and it would cost $2.4 billion to build. More importantly, the plant would be able to capture 3.5 million tons of carbon dioxide each year (equivalent to taking 1 million cars off the road) while generating nearly 600 MW of power (enough for 500,000 homes).[12]

From then on, however, the project became mired in controversy. It suffered unanticipated setbacks, such as too much rain causing a scheduling delay, which increased the overall cost by hundreds of millions of dollars as other steps of the construction process were put on hold. The technology proved too tricky to master, which meant further delays and cost overruns. And, finally, the company was beset by a series of management scandals, including allegations that it hid its failures from regulators – which in turn generated a public backlash, including class-action lawsuits that claimed the costs of those failures had, allegedly, been surreptitiously passed on to consumers paying electricity bills in Mississippi.[13]

The Kemper Project ended up a disaster. Costs ballooned to $7.5 billion and the plant wasn't fully operational even in 2017. By that time, natural gas prices had fallen drastically, and the Kemper plant simply abandoned coal – and its plan to capture CO_2 – and instead started burning natural gas without CCS.[14]

Friedmann, who is married to an accomplished symphony conductor, and whose tenure at the DOE began just as the problems with the project were becoming apparent, compares managing billion-dollar projects to an orchestral performance. 'You need an incredibly high success rate,' he says. 'If 1% of the notes in a symphony are wrong, the audience will find it terrible.'

While management scandals are not under the control of funders like the DOE, the technology failure that compounded the difficulties are not surprising. Technologies that work in the lab or small test plants may not always work when fully scaled. That's why Friedmann says the public money invested in the project was worth it. 'People say "You got $200 million from the government. It should be easy",' he says. 'No, it's like a moonshot. And sometimes moonshots fail.' It's the kind of risk DOE is set up to take, whereby it funds a number of new ideas and fully expects some to fail. Using known technologies makes the job easier, as we'll now see.

About an hour south-west of Houston, Texas, in an area that could cover more than 3,000 football pitches, sits the WA Parish

Generating Station. Its four smokestacks – two taller than the other two – hint at the history of the power plant, which has slowly expanded capacity over the facility's fifty years of operation. With its eight units, four burning natural gas and four coal, it can generate 3,700 MW of electricity – enough to power more than 3 million homes. It's so large it even has its own train line, which regularly brings in more than 15,000 tons of coal from the state of Wyoming, in the north.[15]

It was a typical sunny Texan day in September 2017 when I visited this enormous plant. At all times during my tour, I could see a hillock of coal that covered nearly half the area of the entire complex. You may have seen a power plant at a distance with steam coming out of the cooling towers, but may not have had the opportunity to see it from the inside. The experience gave me a new appreciation of an everyday commodity that's available at the flick of a switch.

Earlier that year, a carbon capture project called Petra Nova (which is Latin for 'new rock') had begun operations, becoming the only second coal power plant fitted with CCS technology. The amount of CO_2 it could capture made it the largest CCS plant of its kind. The US energy company NRG operated it in partnership with Japanese oil company JX Nippon. (The first such plant, called Boundary Dam and run by SaskPower in Canada, began capture operations in 2014.)

'Petra Nova is really five projects in one,' NRG spokesman David Knox told me on the tour: it generates emissions by burning coal, separates carbon dioxide from the exhaust gases, compresses that CO_2 into a liquid, transports it via a pipeline, and finally injects it into the aging West Ranch oilfield.

It cost about $1 billion to build – $190 million of which came from Friedmann's DOE budget – and is capable of capturing and burying 1.6 million tons of carbon dioxide each year. To be sure, CCS was fitted on only one of the four coal units and did away with less than a tenth of the plant's total CO_2 emissions.[16]

Unlike Kemper, however, Petra Nova was built in time and on budget. That's because each of the five steps in the process relied on technologies that had already been mastered by the oil and gas industry. It was simply a matter of putting them together, acquiring the financing, getting all the permissions and then building it – not trivial, but much easier than developing a brand-new technology.

All that happened while market forces were wildly fluctuating, threatening the viability of the project along the way. When the planning for Petra Nova began, in the early 2010s, the price of oil was rising rapidly. The West Ranch oilfield, some 80 miles from the power plant, was generating only about 300 barrels of oil per day, down from 52,000 back in 1970. In theory, the CO_2 captured from Petra Nova into the field would raise its production to 4,000 barrels a day. After taking into account the cost of capturing CO_2, the project would break even as long as oil prices were about $60 per barrel. With oil at $120 per barrel, it was a no-brainer.[17]

But then came the crash in oil prices, just as construction on the project began. The market had misunderstood the supply–demand dynamics, which caused prices to fall to as low as $40 per barrel at one point.

Friedmann thinks Petra Nova wouldn't have survived the crash without a few dogged individuals – like NRG's David Greeson and DOE secretary Ernest Moniz – who stuck it out through the tough years. 'The butts in seats matter,' he says. 'It matters if you have a good project manager inside a company.'

For example, the project had hired Japanese giant Mitsubishi Heavy Industry to build the capture element of the plant. The company needed Japanese banks to provide loans to allow it to participate. DOE seniors reached out to Japanese bank chiefs to assure them the US considered the project a priority. It was a way to assure the banks that NRG, the US partner in the project, had the backing of the US government. 'It's a team sport,' says

Friedmann. 'Everybody has to be on the field. Any substantive climate problem can't be solved by one sector alone.'

But oil prices can't be depended on for a climate solution. When the Covid-19 pandemic hit, oil prices dropped precipitously from $70 per barrel to less than $20. Though they recovered, the average price in 2020 was below $50. That meant NRG had to mothball the Petra Nova project. The world was awash in oil, and there was no financial incentive to capture CO_2 to produce more oil, until prices recovered.

Critics pounced on the news to chalk up yet another failure for CCS. While Friedmann insists that the project's technical success remains untarnished, it's certainly not helping CCS's image that the industry that most needs and can most benefit from the technology isn't committed to building it. After oil prices recovered in 2023, JX Nippon said it would restart the project.[18] That oil prices are one key reason for the start-stop decisions is a reason why proponents say the economics of CCS is yet to become attractive enough to scale.

If you ask an economist how best to tackle climate change, there's a good chance the answer you'll get is put a price on carbon. The idea is that if carbon dioxide causes harm then someone has to pay for it. Right now, that burden is unevenly distributed and does not always track back to the people or organizations putting out the most CO_2 into the air. In 2008 cyclone Nargis killed more than 100,000 people in Myanmar. Unusually warm waters in the Bay of Bengal were found to have made the cyclone much worse than it would have been without climate change.

Economists call these harms 'negative externalities' – a cost of someone's action that is not borne by them. For markets to work efficiently, it's vital that externalities are 'priced in' – that is, actors who cause harm should pay for it. That's one reason why professor Nicholas Stern, an economist at the London School of

Economics, says that climate change is a result of the greatest market failure the world has ever seen.[19]

Simple as it might sound, working out the right price of an externality is fiendishly complicated – hard enough that Yale University economist William Nordhaus won the 2018 Nobel Prize for tackling the challenge and advancing the field.[20] When it comes to climate change, there are difficulties at every step of the causal chain: from understanding how much warming, say, a ton of carbon dioxide can cause to estimating when that warming might lead to harm, and measuring the cost of the harm that will happen at an undetermined time in the future. Despite those difficulties, rough estimates of carbon pricing show how policy can succeed.

A price on carbon can come in many forms. A carbon tax is typically a flat fee a polluter has to pay for emitting a ton of carbon dioxide. For example, Norway's Equinor built CCS projects largely in response to a hefty carbon tax introduced by the country in 1991 – it cost Equinor less to invest in CCS than to pay the taxes it would have incurred otherwise.[21]

An alternative to a straight carbon tax is a cap-and-trade scheme whereby governments allow companies to emit a certain amount of carbon dioxide for free but force those over the cap to buy credits on a carbon market, while companies coming in below the cap are allowed to sell credits on the same market. The cap is slowly lowered in line with climate goals. Partly because it has such a market – also sometimes called an emissions trading scheme (ETS) – Europe has been the continent best able to cut its carbon emissions.

It's also possible to combine the two systems. The UK, when it was part of the European Union, participated in the continent's ETS while also enforcing an additional carbon tax. That made coal much more expensive to burn than natural gas, because coal produces nearly twice the amount of climate pollution as gas does. Now the UK is on the verge of eliminating coal-powered power stations altogether.[22]

Despite the maths, carbon pricing has been the bane of politics. Most countries that have tried have had a tough time getting policies passed. In 2009 the Waxman-Markey bill to create a cap-and-trade emissions scheme failed to find enough support in the US.[23] In 2014 a change of government in Australia led to the repeal of a carbon tax introduced only two years previously.[24] In 2019 Canada's prime minister only just won a re-election bid; his policy to tax carbon pollution was one of the biggest attack points for the opposition.[25] Even outside the Anglosphere, hikes in fuel taxes, an indirect carbon tax, often lead to political revolts – as seen in the *gilet jaunes* (yellow vests) protests in France in 2018.[26]

Friedmann's view is that, while having an explicit carbon price can help, other policies can get CCS going. Tax credits and target quotas for solar and wind power (see Chapters 4 and 9) have helped increase their deployment. The same thing is happening with sales of electric cars (Chapter 3), where many countries are now providing direct subsidies and setting clear targets for car makers. Any of those could also work for CCS.

The Inflation Reduction Act of 2022 – the largest US climate bill – increases incentives for CCS technology, with as much as $85 offered for every ton of CO_2 captured and sunk underground.[27] This is the third time the US government has increased the subsidy it offers for the deployment of CCS – perhaps third time around it will be lucky.

There is a fundamental difference between CCS and other green technologies. In the case of solar, the goal is to make millions and billions of the same type of solar panel. Every unit of production helps the technology become that bit cheaper as engineers learn to improve production. CCS plants aren't modular. Regardless of whether it's being built from scratch or being retrofitted on an existing smokestack, each CCS plant is so large that it ends up being custom built. That means, even as the technology is scaled up, it's unlikely to see the kinds of cost declines that other climate technologies have.

If nothing else works, there's the nuclear option (no pun intended). In the case of coal power plants, mandates to cut sulphur and mercury emissions forced utilities to install equipment that scrubbed those harmful pollutants from exhaust gases. 'We never had a sulphur price or mercury price,' says Friedmann. 'Governments just said you can't emit those anymore. That was policy support.' The same could be done for carbon dioxide.

The difficulty with politics is that, with every election, there's a strong risk of a change in ideology, which may filter through to policy changes. That uncertainty is not conducive to the development of new technologies or businesses.

Policies that have popular appeal across political divides have a greater chance of surviving electoral swings. Despite spending nearly two decades on it, Friedmann had come to conclude that CCS was still lacking wider public acceptance. 'CCS doesn't make anything new. You just don't have emissions,' says Friedmann. 'There's a perception around CCS that it adds a burden, as opposed to it accomplishes a goal. That needs to change.'

Peter Kelemen, a professor of Earth sciences at Columbia University, made the best case for CCS I've ever heard.[28]

In about 1820, Kelemen said, London became the world's largest and arguably most important city. It wasn't just the capital of Great Britain; it was the seat from which the empire's rulers controlled nearly half the world's population. But London, in some ways, was still a backwater – it lacked a central sewerage system. 'If you were poor, you threw your waste down the street,' Kelemen told an audience at the Columbia Global Energy Summit in 2017. 'If you were wealthy, you had a pipe that took it to a cesspool.'

British physician John Snow, now regarded as the father of epidemiology, undertook research in the mid 1800s that eventually showed links between these cesspools and at least three cholera outbreaks, which killed more than 30,000 people in London in the first half of the nineteenth century. To add to the woes, most of the human waste eventually found its way into the River Thames. 'I can certify that the offensive smells, even in that short whiff, have been of a most head-and-stomach-distending nature,' Charles Dickens wrote in a letter to a friend in 1857.

'Then in 1858, there was a summer when it didn't rain,' Kelemen continued. The Thames dried up, and the stench got stronger. It was called the Great Stink. Queen Victoria and the royal court left London; Members of Parliament debated moving to Oxford. Fortunately for those of us who have lived on this Earth since, instead of leaving, they passed legislation to do something. 'They dug up all the largest streets in the world's largest city, and installed central sewers over the next ten years,' Kelemen said. 'It cost about 2% of GDP, and even today it costs about 1% of GDP to maintain the sewers. No one questioned whether that was worth it.'

'Until people have the idea that [throwing CO_2 in the air] is like throwing poop in the street, we're not going to spend what it costs,' he said. In other words, at 2% of global GDP, we can make the CO_2 problem go away. An International Monetary Fund study suggests that GDP in 2100 could be 7% higher if not for climate change.[29] That means the damage climate change can cause far outweighs the cost of preventing it.

Thought about in that way, climate capitalism is about aligning the world with what is considered economic common sense. As much as cutting emissions is going to be hard, it's certainly economically feasible. Government policy plays a huge role in nudging capital to the right places, but it's worth remembering that new policies are experiments that can go wrong. What's most important is to learn from those failures.

I first came to know Friedmann in 2017, a year after he had left the DOE. He was glad of the opportunity to do the work of pushing forward new CCS projects, but wasn't satisfied. With every year, global emissions were rising. 'We are not close to hitting our climate targets,' he blurted. 'We are making progress on renewables, but we are not making progress on CCS.'

We've kept in touch through the years, and, though the world is still nowhere close to building as many CCS plants as are needed to meet climate goals, Friedmann is happy to see there's been some progress. Surprisingly, under Trump, the US government passed a tax credit to support the deployment of CCS, which has bipartisan support in Congress. Across the Atlantic, both the UK and Norway in 2020 announced impressive funding for the creation of new CCS projects that will be built within a decade. Under the European Union's Green Deal, the price of carbon in the bloc's emissions trading scheme is set to rise, which will help support a business model conducive to the success of CCS projects.

Also in 2020, almost all the major European oil companies set goals to reach net-zero emissions by 2050.[30] That goal will require these companies to not just cut oil and gas production but also invest in cleaner technologies – not just renewables but also CCS. Friedmann's conviction remains that carbon management, including carbon capture, is a crucial solution to the climate problem. That's why, after the stint at Columbia, he joined the start-up Carbon Direct, which works on carbon management solutions.

A CCS plant being built in Norway is an exemplary model of the kind of innovation that can come about if governments and oil companies work together. The Northern Lights project will capture emissions from a cement factory near the capital, Oslo, then compress and load carbon dioxide on to a ship to take it up north, where it will be piped offshore and injected underground. Crucially, because the gas can be transported on ships run by oil companies like Shell, Total and Equinor, there is an opportunity for other countries to pay Norway to carry away their carbon waste and store it in the Norwegian shelf – much like some countries pay others to get rid of their household waste.

'If you care about climate change, you got to care about climate math,' says Friedmann. And there's just no way to make the maths work without CCS, not least because humanity has delayed the task of cutting emissions for decades. In fact, keeping to global climate goals will mean even having to draw down some of the carbon dioxide dumped in the atmosphere.

What Friedmann's story shows is that moving an incumbent industry on is hard. But one woman is trying to do just that, by turning her company's business model from being reliant on extracting carbon from the ground to removing it from the air and burying it deep underground.

8

The Reformer

From the sky, the vast flatlands surrounding the city of Midland in Texas look more like an electronic circuit board than planet Earth. It's an endless array of roads that only meet at right angles, like copper cables that conduct electricity, and each block features multiple squares that have been stripped of all vegetation, like resistors and diodes that transform electricity into useful data. The city itself sits in the middle, like a central processing unit, directing the flow and churning out something useful.

The pattern represents something much more primitive than a modern circuit board, however. It's the result of decades spent exploring for and extracting oil. Each vegetation-free square is the scar of a hole punched into the earth that helped drillers reach the layers of rock beneath and draw out the fuel to power vehicles and provide the raw material to make everyday objects like the plastic used to make circuit boards.

Vicki Hollub could reverse the story. The scars will likely remain, but instead of drawing carbon out of the vast Permian basin, she wants to capture the excess carbon dioxide in the air and bury it back underground. Scientists are now certain that such a draw down is necessary if the world is to meet its climate goals. The world doesn't just need to reach zero emissions, it needs to go negative.

Hollub is the chief executive officer of Occidental Petroleum, which has set a goal to become carbon neutral by 2050 and carbon negative after that. The first woman to run a major US oil company, Hollub is attempting to do the impossible: to reinvent an oil company's purpose and thus the industry's image. She says Oxy, as the company is widely known, will then become a carbon management company rather than the carbon extraction company that it is now. 'Our contribution to the world can be a differentiated approach,' she says.[1]

If she succeeds, it'll be quite a change for an industry that has played a crucial role in almost every major global event in the last 100 years. Humanity's utter dependence on securing access to oil – much more so than other commodities – made it central to twentieth-century business, politics and power. That disproportionate importance came with some nasty side effects: it led key players to start wars to ensure control over oilfields; it subjected vast populations to the 'resource curse' that wasted natural wealth; it enabled the installation of despots that then committed crimes against humanity; and it created crony capitalists who sowed doubt about climate science.

Nonetheless, there is no denying oil's role in shaping modern civilization and bestowing it with prosperity – albeit unequally. Despite decades of alarm from environmentalists about its negative impacts, a lack of cheaper, cleaner alternatives has helped oil keep a tight grip on the world. No material comes close to being packed with the amount of energy oil holds in a single barrel. And engineers have used the last century to finely tune the machines that can extract the stored energy with great ease – in engines that are so small they can fit in the palm of a hand or so big that they can move mountains.

Hollub's strategy isn't to abandon oil altogether. But it is an explicit acknowledgement that oil's time at the top of the energy ladder is finally up. Oil will continue to play an important role, especially in some sectors such as aviation, where alternatives aren't available, but it will have an increasingly diminishing role. State-owned oil companies, which today produce the majority of the world's oil, are yet to reckon with this prognosis, but it's only a matter of time as global governments tighten climate constraints.

And Hollub is not alone in this assessment. Most major international oil companies have set goals to drastically cut emissions, with the likes of Shell and BP also having committed to reaching net-zero emissions before 2050. The energy crunch created following Russia's invasion of Ukraine has caused some of these companies to stretch out their time selling oil and gas, but none has abandoned its long-term target.

Reaching that goal will mean reducing oil and gas production or shrinking the company itself until it disappears. In 1990 eight of the top twenty in the Fortune 500 list of the US's largest corporations by revenue were oil companies.[2] In 2023, even in a year that followed high oil prices and helped oil companies to hit record profits, only three remained in this position.[3]

It's not going to be easy changing a 150-year industry that bears a huge responsibility for the warming gases sitting in the atmosphere. But it's also not the first time the industry has had to reckon with changing politics and technology.

The earliest mentions of oil date back to 3000 BCE in the modern-day Middle East. The dark fluid naturally oozed from crevices and people found uses for it: as a remedy for skin conditions, a caulking agent for boats, a lubricant for cart wheels and even as insecticide. In some places petroleum gases escaped and lit 'eternal' flames, giving rise to fire worshippers. Oil was used in warfare, as mentioned in the *Iliad*, Homer's ancient Greek epic poem. By the eighteenth century usage was such that it had become a traded commodity across Asia and Europe.

The age of oil was kickstarted by lighting. Industrialization during the nineteenth century made people in the West richer, and demand for lighting rapidly grew. This was initially met by whale oil, which was extracted from the blubber of sperm whales. But whaling was pushing the species to extinction and driving up the price of the oil. Replacements in the form of animal fat or vegetable fat did a sub-par job, either not burning brightly enough or being foul-smelling. That's when experiments with the black gunk sped up.

Chemically speaking, oil is a complex mixture of chemicals comprised mainly of hydrogen and carbon. Arabian chemists first used heat to separate some of these chemicals in a process that came to be known as refining. Heating caused the chemicals with the fewest number of carbon atoms to reach their boiling point first. Separately collecting and condensing those gases back to liquids created 'fractions' of oil. One of those fractions, which contains chemicals made up of between nine and sixteen carbon atoms, proved to be the most brilliant fuel for lamps.

Initially, the fraction was obtained from a highly viscous form of oil called asphalt. That's why it was named 'kerosene', from the Greek words for wax (*keros*) and oil (*elaion*). But the demand for kerosene quickly began outstripping supply. Among the many people trying to find an alternative source for kerosene was the American entrepreneur George Bissell, who had come across a sample of 'rock oil' that had been dug up in Pennsylvania. After consulting with a Yale University chemist, Bissell became convinced that if he could find enough rock oil then he would be in business and leave behind what had been a life of poverty. Instead of digging it up, however, he would have to drill for it – an idea for which he was widely ridiculed at the time. After a few years of struggle, in 1859 he and his partners struck oil at a depth of 69 feet on a farm in Titusville, Pennsylvania.

Within a few years the population of Titusville grew from a mere 250 residents to 10,000. People came from all over the country to buy up land and set up derricks next to each other. The supply of drilled oil grew so quickly that the price fell from $10 a barrel at the start of 1861 to less than 10 cents at the end of the year. The low cost, however, also helped oil capture its first market away from the competition. Refiners that were using asphalt to make kerosene quickly converted to exclusively using crude oil.[4]

That first boom-and-bust cycle, however, soon became a feature of the oil industry. Whenever there was a shortage of supply and prices rose, drillers would scour the planet in search of new oilfields. As soon as one was found, others would rush to extract as much as they could, in turn flooding the market and pushing prices down. Until, that is, demand caught up and prices rose again, which would then trigger the next boom-and-bust cycle.

The booms could be spectacular, and this made the busts more palatable. Famously, one oil well drilled in the 1860s led to a return of $15,000 for every dollar invested in a span of less than two years. It also meant that lots of people tried to game the market.[5]

Under the banner of Standard Oil, founded in 1870, John D. Rockefeller built one of the world's largest corporations – forty-one years later it was deemed an illegal monopoly. The broken-up entities, some of which we know today as ExxonMobil, Chevron, Marathon and ConocoPhillips, proved to be even more valuable than the monopoly was, making Rockefeller one of the richest people in modern history.

Rockefeller left a deep mark on the oil industry. Despite the irregularities that brought about the downfall of Standard Oil, three lessons from his time managing the company have become core tenets of how the industry operates. Some of those lessons may seem obvious today, but in the nineteenth century they were revolutionary. And they may also make it harder for oil companies to transition to providing clean fuels in the twenty-first century.

First, economies of scale help. Standard Oil started out as a company that refined oil someone else had drilled and sold kerosene to wholesalers who then resold it to retail customers. Slowly, however, Rockefeller recognized that one way to reduce risks and increase profits was to get good at all parts of the business, from extraction to retail sales. This strategy, known as vertical integration, helped him hoard crude oil or refined products when prices were low and then sell them when prices rose – thus maximizing profits. Creating a brand for his company helped differentiate Standard Oil in a business that was selling a commodity.

Second, hold on to cash to ride out the boom-and-bust cycle. When oil prices fell, Rockefeller looked for opportunities to invest cash – everything from buying up failing firms to exploring for new oil. He knew that a boom was coming and he used the cash he held to ready himself to make the most of when oil prices inevitably rose again.

Third, create new markets. Rockefeller learned that making a product people want is not always enough. With so much oil available, it is better find new markets for the product. Over its lifetime, Standard Oil expanded its operations to supply oil and its products to Asia and Europe as well. When oil's consumption via kerosene began to top out, after Thomas Edison popularized the light bulb, Standard Oil found a new customer in petrol-powered cars.

The oil moguls that followed, whether chief executives of publicly listed companies or monarchs of oil-producing countries, have adapted Rockefeller's lessons for the new challenges the industry has faced as the world has globalized and financing has become more sophisticated. Those lessons have guided the oil industry for more than 100 years – a period in which oil has gone from being a novelty to providing for the largest share of global energy demand.

Vicki Hollub was born in 1960 in a suburb of Birmingham, Alabama, just as oil was displacing coal at the top of the energy charts. She studied mineral engineering at the University of Alabama and joined a small oil company, Cities Service, in 1981.

The very next year Occidental Petroleum bought Cities Service and Hollub has remained at Oxy ever since.

The 1980s were Oxy's seventh decade as a company and the start of its most explosive growth period. Soon Hollub moved from technical roles to managerial ones with assignments abroad. She worked in Russia just after the fall of the Soviet Union; in Venezuela with military escorts keeping her safe from guerrillas active in nearby Colombia; and in the Ecuadorian Amazon rainforest.

When she joined Oxy there was only one other female engineer and a handful of women working as geologists. She says that she was often the only woman at an oilfield, where she fitted in by talking American football with co-workers. Even today, less than a quarter of the oil industry is female and barely 2% of executives are women. That share is even lower in the energy industry in China or India.

Hollub credits her boss at the time, Glenn Vangolen, for backing her. 'He took a chance,' she said. 'Even though we were generally open to diversity, he did have some opposition and he had to stand up and say, "She is going to do that job. She is going to do it well."'[6]

The turning point in Hollub's career as a leader came in the 2000s, after Oxy bought Altura Energy.[7] The acquisition gave Oxy access to the largest oilfield in Texas and expertise in injecting carbon dioxide underground as a means to extract more oil (see Chapter 7). Altura's fields would also play a crucial part in the coming fracking boom, which would unlock new oil reserves trapped in shale rock through a mixture of new drilling technologies. Her ability to integrate Altura's assets into Oxy won her plaudits, and a series of promotions followed. They happened to come at a tumultuous time for the company.

For six decades only two chief executives had led the company. First, Armand Hammer, who took over in 1957, and then Ray Irani, from 1990. Both oil tycoons were colourful characters boasting lavish lifestyles, big pay packages and vast private art. Hammer was alleged to have been a Russian spy and Irani became one of the highest-paid oil CEOs in modern history.

By the end of Irani's tenure in 2011, however, Oxy's expansionist streak combined with a volatility in oil prices meant the company's balance sheets were in tatters. Oxy needed a different kind of management style, and Hollub's down-to-earth approach appealed to the company's board. It's also when Hollub realized the company needed more carbon dioxide if it were to keep its assets producing oil at a profitable rate.

Between 2009 and 2016 Hollub was promoted five times, with the final promotion being to the top job as CEO. She became the first woman to lead a large Western oil company, and her rise in an industry dominated by men remains exceptional. Two years later Oxy published its first climate report, laying out the kind of shift the oil giant would make under her leadership. The report also marked the close of one of the darkest periods in the history of the oil industry, bringing to an end decades of efforts to sow doubt about climate science.

Historians have found that the oil industry knew about the link between fossil fuel burning and global warming as far back as 1959.[8] They have also found evidence that industry groups attempted to sow doubt about climate science as far back as 1980. The most well-recognized push, however, began in 1989 with the formation of the Global Climate Coalition (GCC), only months after the United Nations had established the Intergovernmental Panel on Climate Change.[9]

At its height the GCC counted most major coal and oil companies as members, alongside powerful industry groups such as the American Petroleum Institute and the US Chamber of Commerce. Its main goal was to oppose regulations to mitigate climate change, and it did that by adopting the playbook the tobacco industry had used in the 1950s to cast doubt on the link between smoking and lung cancer. The idea was to use disinformation campaigns through advertising and highly paid lobbyists to cast doubt on the certainty of the link between fossil fuel use and rising global temperatures. Though decades apart, both campaigns even used some of the same public relations firms.

The GCC was disbanded in 2001 and the IPCC still puts out alarming scientific reports every few years, which tells you who got the facts right. But in its twelve-year existence, the GCC managed to do a lot of harm. It took advantage of the IPCC's open process to create baseless scandals by attacking the credibility of scientists who authored the reports. It paid for 'experts' to turn scientific facts into debates. The GCC and its members also published advertisements that questioned the science. Taken together, the GCC's work lived up to the infamous tactic from the tobacco playbook: doubt is our product since it is the best means of competing with the 'body of facts' that exists in the mind of the general public.

The most tangible result of the GCC's work was that the US never ratified the Kyoto Protocol, which was meant to legally bind rich countries to cut emissions and lay the ground for more ambitious global agreements on tackling emissions. The more insidious outcome has been the creation of a much larger and more dispersed climate denial machinery that continues to operate and delay action on climate change.

The largest companies in the GCC were the oil majors. Between 1986 and 2015 five oil companies we now know as ExxonMobil, BP, Chevron, Shell and ConocoPhillips spent $3.6 billion on advertising and a further $2 billion on government lobbying. Experts say that these figures from publicly listed companies only represent the tip of the iceberg, with spending on trade associations, political campaign donations and so-called dark money groups still largely hidden.[10]

Even though all oil companies have since accepted the overwhelming science of climate change and the role fossil fuels play in it, many haven't quite accepted the changes that will be needed to truly tackle the problem. Take ExxonMobil as an example. Like all oil companies, it was forced to cut its spending in 2020 in the face of the economic recession brought on by the Covid-19 pandemic. As a result, it chose to put on hold a carbon capture facility in Wyoming that was to cost $260 million, but remained steadfast with its $9 billion investment in Guyana that would add billions of barrels of oil to the world market.[11]

These kinds of actions show that many in the oil industry are still not serious about taking the actions needed to cut emissions. They make many doubt whether any oil company will truly ever do enough to keep a profitable product in the ground. Despite that, perhaps profits are exactly what make Hollub's change of religion – from extracting carbon out of the ground to putting it back in – more believable and lead many to think her plan may actually come to fruition.

Hollub was named CEO in the same year the Paris Agreement was signed. The world for the first time accepted the urgency of addressing the climate crisis and a new era in global politics was born.

In 2018 the Pope invited Hollub and a few other leaders in the energy industry to the Vatican for a conference called Energy Transition and Care for Our Common Home. At the conclusion of the meeting the Pope addressed the group directly: 'This is a challenge of epochal proportions,' he said. 'Civilization requires energy, but energy use must not destroy civilization.' He called on Hollub and her kin to use their 'skills in the service of two great needs in today's world: the care of the poor and the environment'.[12]

When she returned, something had changed. 'The meeting materially affected her,' said a close collaborator of Hollub's. 'She came back from Rome and thought differently about her role as a CEO and the value proposition of Oxy to the world.'

Hollub expected that the world would continue to consume oil for decades to come, but it could not have the greenhouse gas impact that it was currently producing. Oxy was the world's leading company when it came to injecting carbon dioxide to increase oil production. But that carbon dioxide was currently being mined from other underground fields, even as it was an excess of the gas in the atmosphere that was becoming a problem. She realized that she could kill two birds with one stone, if she stopped mining for carbon dioxide in underground fields and instead mined it from the air. And, as it happened, there was a Canadian start-up called Carbon Engineering that was looking to scale up that very technology, called direct air capture.

Here's how a direct air capture plant works. Large fans suck huge volumes of air and pass it over corrugated sheets. The sheets are coated with a solution that reacts with carbon dioxide in the air, leaving behind a carbon-rich solution. That solution is then brought into contact with quicklime (or calcium oxide), which reacts to form pellets of calcium carbonate. These pellets are then heated to about 1,000 °C to release carbon dioxide as a pure stream of gas and recreate quicklime, which can then be used for another cycle to capture more carbon dioxide from the air. The pure stream of greenhouse gas is then injected deep underground and kept out of the atmosphere.

Within months of meeting the Pope, Oxy made an investment in Carbon Engineering. And by 2019 Hollub had a plan for scaling up its technology. Oxy would build a Carbon Engineering plant in its Texas oilfields that would capture and store as much as 1 million metric tons of carbon dioxide from the air and be classed as negative emissions. If she injected more carbon dioxide in the fields than was currently being extracted, the accounting could mean oil coming from those fields could be considered 'carbon negative'.

The process behind direct air capture is similar to conventional carbon capture except that the concentration of carbon dioxide in the two streams is vastly different. The exhaust of a coal power plant, for example, contains about 10% carbon dioxide. The atmosphere, on the other hand, is made up of merely 0.04% carbon dioxide. And just as it gets harder to search for a red M&M in a sea of blue M&Ms, so the process of capturing carbon dioxide from a very dilute stream is much harder than capturing it from a concentrated stream.

The main difficulty Hollub faced was that the plant would cost hundreds of millions of dollars to build and then hundreds of millions each year to keep running. As a first-of-its-kind plant, the cost of capturing each ton of carbon dioxide was expected to be in excess of $200. That, in turn, would raise the cost of each barrel of carbon-negative oil by many tens of dollars, relative to a conventional barrel. Who would pay that premium?

Only months before Oxy announced its plan to build the direct air capture plant, a new rule in California had created a market for carbon-negative oil. If all went to plan, Oxy might even make a profit from building the very first large-scale direct air capture plant.

Among US states, California has long stood out for its environmental leadership. And within its government departments that have led on climate, the California Air Resources Board (CARB) deserves outsized credit for enabling that leadership.

For much of its existence since 1967, CARB was focused on cutting air pollution in California's rapidly growing cities. In the 2000s, as the climate alarm bells became louder, it was given the mandate to set greenhouse gas regulations.

With transportation contributing to the biggest share of California's emissions, CARB recognized that there was no way to meet the state's climate goals without drastically cutting transport emissions. Its answer to the problem was the creation of the low-carbon fuel standard (LCFS), a cap-and-trade programme whereby CARB sets a cap on the emissions from the transport sector and brokers trade credits for each ton of carbon dioxide reduced using an alternative in the form of hydrogen, biofuels or electric cars.

Trading of LCFS credits started in 2011. The cost of each ton of carbon dioxide avoided through the cap-and-trade programme has been steadily rising through the years, as CARB reduces the cap. In 2020, for example, credits traded for an average price of $200 per ton.

The success of the programme led to its expansion. In 2018 California committed to reaching net-zero emissions by 2045. CARB realized that, even as the number of non-petrol cars was growing, emissions from aviation were not trending in the right direction, however, and in 2019 it duly allowed the use of direct air capture as part of its LCFS trading.[13]

In other words, by the time it made its announcement, Oxy felt quite confident that it could get as much as $200 for each ton of the carbon dioxide it was going to capture. That would write off the expense it would incur in the process of extracting the gas from the air. Better still, a separate US federal tax credit for burying carbon dioxide and aiding oil extraction would bring in another $35 per ton. Combined, the project would be in the green.

The plant's construction began in 2023, and it was more good news. The Inflation Reduction Act passed in the previous year provides additional incentives for direct air capture.[14] That's yet another revenue stream that Hollub can tap. The bill also increased the sum companies can get for burying CO_2 to extract more oil. At the same time, corporations with net-zero goals are also increasingly looking to bury carbon dioxide that they have to otherwise emit through air travel for their executives.

If everything goes to plan, the plant will be in operation by 2025 and it will be a rare case of an emissions-cutting technology generating profits from the very first large-scale plant. Hollub says that she can't keep up with demand for carbon removal and that's why Oxy is now planning to build seventy-five such plants.[15]

As the oil industry moves away from the business of extracting oil and gas, it is having to reinvent itself. In getting to this point, Hollub has adapted Rockefeller's key lessons, that economies of scale help, deploying cash carefully enables you to ride out difficult times, and the importance of finding new markets. Oxy is scaling up direct air capture to bring down technology costs, it is using the cash from current oil production to support the transition, and it is finding new sources of revenue through burying carbon instead.

This transition is happening across the world. Norwegian firm Equinor has become a big player in offshore wind. British giant BP is winding down its oil and gas production, and using the cash to grow renewables and EV charging networks. French giant Total is taking a BP-like path, but is also investing in lithium-ion batteries.

But it's a messy transition. The oil and gas behemoth Shell bought utility companies a few years ago in a bid to become a provider of clean power, but in early 2023 reports suggested that it is considering divesting.[16] After record profits in 2022, BP also said it will reduce fossil fuel production more slowly than it first planned. At the same time, US company ExxonMobil, which had resisted making any real moves to diversify from fossil fuels, organized its first Low Carbon Solutions Spotlight in early 2023.

There is no guarantee that these oil companies will succeed. For the last century they have developed and refined an expertise for dealing with two types of energy commodity: oil and gas. The transition means competing with companies that have had a longer history running a utility, building solar or wind farms and manufacturing batteries – all of which need different skill sets. As climate capitalism becomes a bigger force, companies that focus too much on short-term profits will be unable to survive for very long.

That leaves the door open for further disruption. Recognizing the need for negative emissions, a new class of start-ups is rising up to the challenge. In 2021 Elon Musk announced a $100 million prize for ideas that could capture carbon dioxide from the air.[17] In 2022 big global companies came together and created an almost $1 billion fund to support carbon-removal technologies.[18] It won't be just tree-like machines, but other creative ideas, from growing sea forests to carbon farming, will also be on the table.

There is one route that these companies can take, which is to move entirely to clean energy. No major oil company has yet completed that transition, but the experience of a small European oil company shows it can be achieved in as little as fifteen years. That company, once called Danish Oil and Natural Gas, is now much more famous globally as the offshore-wind giant Ørsted.

9

The Enforcer

Ørsted is a poster child of the energy transition. Once a corner-stone of Denmark's oil and gas business, the company has turned itself into a wind power behemoth. In less than two decades it will have eliminated almost all of its emissions. That's why Ørsted's story – from Danish Oil and Natural Gas (DONG) to a clean energy giant – has been fodder for case studies in business schools around the world, showing how careful planning and foresight can be brought to bear in changing times.[1] Politicians point to Ørsted to nudge companies to follow the green path to riches and create the jobs of the future.[2] Climate activists use it as an example to call out the excuses that large oil companies make about the difficulty of transitioning to cleaner sources.

All that being said, the credit for Ørsted's success should go as much to government policies and well-timed subsidies as it should to corporate leadership. While Ørsted's transformation can provide a blueprint showing what fossil fuel companies ought to do in the coming decades, it's an incomplete plan without understanding the decades-long energy history of Denmark that created space for something like Ørsted to form and thrive. That story starts, as many energy transition tales do, with an oil crisis.

The 1973 oil crisis was a major event for most rich Western econ-omies. But no country was likely more transformed by it than Denmark. Initially, the Nordic country wasn't in the crosshairs of

Saudi Arabia and its allies when they announced an oil embargo in October 1973. Then in November, at a closed-door meeting, the Danish prime minister made remarks in support of Israel in its war against Arab countries.[3] As a result, the price of oil jumped more than 300%, with a genuine risk of entire countries running out of fuel.

Denmark was hard hit. At the time, 90% of the country's energy came from burning oil and almost all of that oil came from the Middle East. After the embargo was imposed, Denmark pursued many similar policies to other oil-dependent Western countries to reduce oil consumption: lowering speed limits, turning off every other street lamp, banning the use of cars on Sundays, promoting shorter and colder showers, and so on. That all helped a bit to deal with the immediate crisis, but the shock left a deep impression on citizens and the country's political and business class.

'It was a very dramatic wake-up call for Danish society,' said Anders Eldrup, who had then just graduated from university and begun working for the national government in the capital city of Copenhagen. 'Politicians said "never again".' Eldrup would go on to become the chief executive officer of DONG.

The next decade saw a series of government-led deep reforms, supported by private industry and aimed at diversifying the sources of energy supply. What followed can now be understood as a six-step plan:

1. Convert power plants that burn oil to burn coal. While most of the oil came from Arab countries, coal could be imported from many places, including from allies like the United States.
2. Extract your own oil and gas. Recent discoveries in the Danish North Sea made that viable.
3. Reduce energy use overall through energy efficiency measures. Fuel not used is fuel saved.

4. Use the extra heat from power plants for heating homes. That bumps up how much usable energy can be extracted from the power plant without increasing the amount of fuel burned.

5. Turn waste from homes and industry into energy. Instead of waste going to landfill, burning it displaces the need for fuel.

6. Find ways to use the abundant wind that Denmark is blessed with. That would have to start with developing the necessary technologies.

Even though climate change wasn't on a list of top worries at the time, it turns out that many of the steps Denmark took at this point are exactly what countries around the world need to be taking today, to reduce their dependence on fossil fuels and cut emissions.

Let's first look at how the government implemented those reforms. Danish leaders had been worried about the country's utter dependence on imported energy for some time before the oil crisis hit. In 1972 the government created separate companies to exploit the country's newly found oil and gas in the North Sea. After the start of the oil crisis they merged to form DONG, which was to supercharge existing efforts.

Next came the creation of the Danish Energy Agency (DEA) in 1976. The energy system is complex and ever-changing, and the DEA was set up to advise the government on policy that would work across different sources of energy and its many types of consumer. It started by promoting the development of wind power and providing subsidies for the deployment of modern turbines.[4]

While neighbouring Sweden embraced the new and emerging form of fossil-free energy in nuclear power, Denmark was strongly opposed to it. That led the government to ban the building of nuclear power plants in 1985. With oil also out of favour, it had to focus on maximizing the fuels it could still access: coal, wind, natural gas, biomass and waste.

Beyond the immediate measures the government took in 1973 to cut oil use, the longer-term reduction in demand for the fuel came after a series of laws and executive orders pushed for more efficient heat supplies. A little less than half of the country's energy consumption happened in homes and commercial buildings, where most of the time oil was burned to meet heating needs.

The Heat Supply Law of 1979 required local municipalities to map out existing demand for heat and create a plan for meeting that need via the use of natural gas at home or through much more efficient district heating systems that may use any fuel except oil. The government also changed the planning rules to ensure that any new infrastructure would use the district heating system or natural gas for heating.

A district heating system involves more efficient burning of fuel in a centralized furnace and then transporting heat, typically as hot water, into homes. It provides two advantages. First, as cleaner sources of fuel become available, the furnace can be tweaked or changed to use that fuel, rather than having to change individual boilers in hundreds of thousands of homes. Second, the centralized furnace could be part of a co-generation plant that produces not just heat but also electricity. The most efficient natural gas power plant, for example, can convert less than half of the energy content of the gas into electricity, with the rest of the unusable energy dumped into the atmosphere as waste heat. A co-gen plant turns a lot of that waste heat into usable energy and thus boosts the efficiency of the power plant to 90%, which means less fuel has to be burned for the same benefit.

Such changes to a country's energy system are typically expensive. Many of the small district heating plants were owned and operated by local municipalities, which provided energy to citizens without turning a profit. Those municipalities were offered flexibility in the fuel they used, which meant they could use whatever was cheapest – as long as the choice was not oil. Still, some of the biggest investments needed, such as a natural gas transmission line from the North Sea and gas pipes to homes, were funded by the government. Some of those funds came from taxes on energy use and eventually emissions, first on oil and electricity (both in 1977), then coal (1982), carbon dioxide emissions (1992) and gas (1996).

The result? Between 1972 and 1990 the share of buildings using natural gas increased from 0 to 10% and those using district heating increased from 20 to 40%.[5]

This is a short version of the history and it might look straightforward, but that wasn't the case. 'It was not achieved with one fully elaborated plan,' says Eldrup, 'but arrived at through trial and error.'

This series of government reforms provided new opportunities for private companies.

Consider the example of energy efficiency. The International Energy Agency calls efficiency the 'first fuel – the fuel you do not have to use – and in terms of supply, it is abundantly available and cheap to extract'.[6] Put more simply, energy efficiency can mean the same comfort but with lower energy bills.

While lower energy costs are attractive, politicians have struggled to rally interest in energy efficiency. It's easier to sell building a new shiny solar power plant or wind farm. It's harder to sell insulation and electrical appliances that have higher efficiency ratings but look the same as less efficient, cheaper ones.

Done right, government policies can work. For example, taxes on energy incentivize people to use less of it and mandates for efficiency measures in codes used to approve new buildings or appliances ensure people use less energy to start with. Denmark introduced just the right number of policies to spur industry to step up. 'We had some success with reducing energy consumption,' said Eldrup. 'We also had business success.'

Four Danish companies – Grundfos, Danfoss, Velux and Rockwool, all founded in the first half of the twentieth century – had energy efficiency solutions ready when the 1973 oil crisis hit. Grundfos made energy efficient pumps needed to move water and ventilate buildings. Danfoss made radiator valves that helped reduce energy use in heating. Velux offered windows that let in heat but did not let it escape. And Rockwool made insulation material that lowered the amount of energy needed to keep a building warm. These companies did not just make buildings in Denmark more energy efficient, but also went on to capture a significant share of the market abroad.

Denmark's energy efficiency improvements have made a big impact over the decades. In 2019 the country consumed about 20% less energy in total than it did in 1973, even as its population has grown by 20% and its economy has doubled in size. For comparison, over the same period France and Sweden, which also saw their economies and populations grow substantially, consumed 24% and 18% more energy respectively. Germany and the UK had better figures to show, with 9% and 19% lower energy use respectively, but they still couldn't beat Denmark.*

'Many countries cut energy use, but we did it more successfully because we had this tough experience,' says Eldrup. A tangled web of government policies, changing business conditions and entrepreneurial spirit can combine to create green champions. Ørsted is probably the best example of how it happens. But before we get to its story, we need to look at how the Danes mastered wind power.

*

---

* Calculations based on data from BP's Statistical Review of World Energy 2021.

In the middle of the government's energy reforms of the 1970s, some Danish citizens took the energy question to heart and some even tried to find solutions on their own. In 1978, students of Tvind high school in Jutland worked with teachers and other experts to build the world's largest wind turbine at the time. Tvindkraft, as it's called, was built to show that Danes could harness wind power and as a way of countering advocates of nuclear power. That committed citizens could achieve such a feat also made it seem like tapping wind power wasn't so hard.

The turbine is generating electricity even today, and it holds the record of being the world's longest-operating wind turbine. Its tower is 53 metres tall and the diameter of the wings is 54 metres. It's capable of producing just under 900 kW of power, which is sufficient to power 500 homes. One of Tvindkraft's greatest legacies is probably inspiring Henrik Stiesdal, who some call the godfather of modern wind power.[7]

In 1976 Stiesdal had just finished high school and was due in a few months to join the military for mandatory conscription, which can last from a few months to a year. His father took him to visit Tvind during the construction of the turbine, and he was immediately inspired by watching others his age involved in the process. At the site he found books on renewable energy and wind power.

Having just finished a course in physics and with a few months before his service was due to start, Stiesdal put his limited knowledge of wind turbines to use. The first turbine he built had two blades and was small enough that he could hold it in his hands. 'Once you got it spinning, it got to be alive, and you could feel all the small things in the wind,' Stiesdal recalls. 'I was hooked.' He spent months, bar the break for military service, building turbines, starting with a two-blade version and then updating it to the three-blade version that we see the world over today.

By 1978 he had joined the Danmarks Vindmølleforening (Danish Wind Turbine Association), which had fifty other like-minded tinkerers. During their experiments he met a local machinist named Karl Erik Jørgensen and found in him a partner with complementary skills. Together, they went hunting for scrap metal and cheap used equipment. They sought money from wherever they could find it to keep building better wind turbines. The biggest success came when they won funding from the Danish Technological Institute's Inventors' Bureau.

In mere months Stiesdal and Jørgensen had erected turbines that could power a small village. While building one of those, Stiesdal almost died. As he wrote in *Wind Power for the World* in 2013:

> At that time, in 1979, it was not ordinary among pioneers to use personal safety gear such as safety harnesses. At a certain stage of installation, I was standing at the top of the tower on the platform, mounting the belt that would connect the small generator to the big generator. By a mistake, the small generator was suddenly powered up while my hands were still on the belt. As a pure reflex without thinking where I was standing, I jumped back. And it was only because my shirt was partly caught by a fitting that I did not fall down from the height of 18 meters. With a torn shirt and my heart in my mouth, I was able to climb back onto the platform having acquired a new bit of experience.[8]

Within months of having completed this turbine, engineers from a local manufacturer of cranes came over to visit. The Danish government was subsidizing the development and deployment of wind power. The company thought they could turn Stiesdal's design into commercial success, and they signed a licensing agreement. That company was Vestas, which today is the world's largest maker of wind turbines.

Since then Stiesdal has remained a wind engineer with hundreds of patents to his name. After conquering onshore turbines, he has gone on to build designs for offshore wind turbines and even floating ones. He is still active. Sadly, Jørgensen did not see the level of success their work achieved. He died in 1982 from cancer.

After 1984 the Danish government introduced a so-called feed-in tariff that guaranteed a certain price for wind power regardless of the grid price. That created a new market for small wind turbines and helped companies like Vestas to grow. In 1987 the Danish power company Elkraft built a 3.75 MW wind farm that was the largest in Europe at the time.

The development of offshore wind power was crucial for Denmark. The country didn't have a lot of land, but it did have access to a lot of sea. Crucially, it had access to areas of the continental shelf where the sea floor was shallow enough to build a solid base to erect offshore wind turbines.

In 1991 Elkraft built the world's first offshore wind farm, Vindeby, made up of eleven wind turbines that could each generate 450 kW. In 1995 another power company, Elsam, built a similar-sized offshore wind farm, Tuno Knob, with ten turbines that can produce 500 kW of power each. Both were built because the government mandated power companies to build offshore wind farms as part of its support for clean-energy innovation. The wind farms would go on to become part of DONG and then Ørsted.

Before 2005 DONG remained a state-owned entity that managed the exploration and production of oil and gas from the North Sea and the distribution of gas across Denmark. However, the European Union was looking at liberalizing the energy market. That meant other European companies could start to compete for the businesses that DONG had a monopoly over.

Eldrup, who had worked in the Danish ministry of finance for nearly two decades, was appointed the chief executive officer of DONG in 2001. He looked at DONG's assets and realized that while the company had contracts extending out a decade to supply gas, it did not have much of a future beyond that. 'It was sort of a dead end,' he says. He realized that the only way DONG could compete with European energy companies like Sweden's Vattenfall and Germany's RWE was if it wasn't just a gas company or a utility but did a bit of everything.

So DONG embarked on an acquisition spree, which is something the government supported in order to protect Danish interests in the energy sector. By 2006 DONG had merged with six power companies to form DONG Energy. Now it had not just oil and gas exploration along with gas distribution, but also coal power plants and customers directly paying for electricity.

Elkraft and Elsam were part of the acquisitions, which meant DONG Energy now also owned three offshore wind farms: two from Elkraft and Elsam plus one that DONG had begun building in 2005. Though a small part of its total revenue, DONG Energy had suddenly become the largest player in the world in the offshore wind market.

'Our thinking was we are now in a safe place,' said Eldrup. DONG Energy had assets across the whole chain: production, transmission and retail.

Soon, however, another crisis came knocking. The European Union's emissions trading system had been launched in 2005 and it covered the power sector. As the price of carbon rose, so coal power started becoming less profitable. At about the same time Copenhagen was named host of the decisive 2009 COP meeting. Governments of host countries to a COP meeting are typically expected to push for higher climate ambition. In support of the government's goals, Eldrup set DONG Energy a goal to switch from 85% fossil fuels and 15% renewables to 85% renewables and 15% fossil fuels by 2040.

While making that promise, however, DONG Energy was also planning to build a new coal power plant in Greifswald, northeast Germany. That seemed to contradict the priorities of a government that was still the majority shareholder in the company. Environmental activists took notice and DONG Energy had to deal with many protests. 'We had succeeded in acquiring all these coal power plants, but maybe it was not so good after all,' Eldrup said. The global leadership on offshore wind started to look a lot more attractive. DONG Energy was forced to cancel the new coal power plant at a considerable cost, in the order of tens of millions of dollars.

The COP 2009 meeting failed to produce the outcome that many had hoped for. The world had to wait till the COP 2015 meeting in Paris for it. Still, DONG Energy maintained its climate goals and began to build more offshore wind farms, especially in the UK.

Two things made doubling down on offshore wind feasible. First, the company had expertise in offshore work exploring for oil and gas. It could redeploy those skills in the wind business, alongside the existing knowledge base its staff had in building the world's first few offshore wind farms. Second, according to government rules, the company was not allowed to exploit oil and gas resources outside Danish waters. However, Eldrup said, there was no such rule on exploiting wind resources. The UK government was offering attractive contracts with a guaranteed price of electricity for offshore wind. DONG Energy was the most mature offshore player and it had the pick of the choices when it came to attractive locations for the wind farms.

By 2012 DONG Energy was in deep trouble. Gas prices began to drop because the US had found domestic supplies through fracking. That caused the Danish giant to record huge losses. It still had a lot of debt following the acquisitions. The combination led to the downgrading of DONG Energy's credit rating.

In a debt crisis, companies are forced to focus. DONG Energy had to put a lot of assets up for sale to ensure it could pay down its debt. There was a reshuffle of the company's leadership, with Henrik Poulson replacing Eldrup. In 2013 DONG Energy got a

$2 billion investment from Goldman Sachs and Danish pension funds to shore up its finances.[9] By 2014 the energy giant was past its biggest troubles, and was left with only assets tied to offshore wind alongside oil and gas. The process of decarbonizing had sped up, but not in a way Eldrup had imagined.

By 2016 Poulson turned DONG Energy's balance sheet around. The company continued to maintain its lead as the world's largest offshore wind farm company, having constructed nearly a quarter of the global capacity by then. Investors were starting to see the benefits of the transformation and that spurred the company to list on the stock exchange. At a valuation of $15 billion, it became Europe's largest initial public offering of the year. Goldman Sachs and the Danish pension funds made a handsome return on their bets.

In 2017 Poulsen began the process of selling off DONG Energy's oil and gas assets. North Sea oil and gas were becoming expensive having been exploited for decades. At the same time, offshore wind was getting cheaper to build. That convinced Poulsen to turn the company into a pure-play renewables company.[10]

By the end of 2017, having divested itself of most of its fossil fuel assets, DONG Energy changed its name to Ørsted. The name was in honour of Hans Christian Ørsted, the Danish scientist who discovered that electric currents create magnetic fields and thus became the first to find the connection between electricity and magnetism. Wind turbines generate electricity using the very principle that Ørsted had discovered. The moving blades turn a rotor inside the generator with a magnetic field and that results in the movement of electrons to power so many things around us.

DONG's transformation into Ørsted isn't quite the textbook MBA case study. Instead, it is the result of a mixture in equal parts of accident, timing, policy and entrepreneurialism. And it happened, not over years, but decades. That's not the kind of luxury available under tighter climate deadlines. However, through trial and error, there is a reasonable formula available for corporations looking to change their business model: choose a technology to back, work with government to create the policies that help deployment, and then use the expertise to find new markets. Commercial opportunities might have been a big enough reason for Ørsted's transformation, but it certainly helped that the government was supportive.

The government remains Ørsted's largest shareholder with 50.1% of the shares. In 2014 Denmark passed a climate law that legally bound the country to become a low-carbon society by 2050. With the country committed to cutting emissions in line with what science demanded, Ørsted's green transformation was a perfect fit with national priorities. These kinds of laws are starting to permanently change how businesses operate, and the UK is perhaps the furthest ahead in showing the world how to use the power of laws to change the flow of capital.

10

The Campaigner

One of the most tumultuous days in the history of Britain was 23 June 2016. On that day the country voted by a narrow majority to leave the European Union. The pound slid to a thirty-year low, the FTSE 100 benchmark stock index fell by more than 10%, London's finance industry was in turmoil, and businesses across the country were in a state of shock. Faced with an impending Brexit he'd campaigned against, the then prime minister, David Cameron, resigned, which triggered a leadership contest in the Conservative Party. The opposition didn't fare well either. Jeremy Corbyn, who led the Labour Party, clearly couldn't convince the British public to vote against Brexit and found himself losing a no-confidence vote of his party's Members of Parliament.

Despite the economic and political chaos, the UK government did pull itself together a week later to pass a historic piece of climate legislation. On 30 June it agreed to set a new climate target of reducing emissions by 57% by 2030 relative to 1990 levels. It was the most ambitious climate target yet announced among all the world's large economies.

It wasn't the last time political upheaval in the UK would lead to a more progressive approach to climate change. Three years later, Cameron's successor, Theresa May, fell victim to similar political bickering about Brexit and, in May 2019, announced she would step down as prime minister. Unrelated to any of those concerns, in June 2019, as one of her last legislative moves before leaving the job, she and her government agreed to commit the UK to reaching net-zero emissions by 2050, upgrading a previous commitment of reducing emissions by 80% by the same date.

May's successor Boris Johnson was not immune to the pressure to do more on climate change, even as his government struggled through the tiring negotiations needed to actually leave the EU and faced the devastating Covid-19 pandemic. In the lead-up to hosting the United Nations' annual climate conference COP26 in 2021, he had to show that the UK was serious about its intent to act on climate change. That led Johnson to upgrade the country's climate goal to reducing emissions to 68% by 2030 relative to 1990 levels. He went further, setting a goal of 78% reduction by 2035.

Put another way, within a span of five years that saw three different prime ministers, the UK had pushed the target date of reducing emissions 80% a full fifteen years – from 2050 to 2035. One of the main reasons behind this upward spiral of ambition on climate change is the power of laws.

'If you set up the right institutions and the right framework, you can still make progress,' says Bryony Worthington, a member of the UK parliament's House of Lords. She would know because she was a lead author of the Climate Change Act of 2008, which established the institutions and framework for the target-setting spree that has followed. Parliament passed the law with 463 votes in favour and only three against, with all eight political parties showing overwhelming support.[1]

Worthington believes that legal infrastructure, and its strong base of support, will successfully position the UK in its efforts to become a carbon-free economy – regardless of which political parties end up in power in the coming decades. She also thinks it's quite likely the UK will go beyond reaching net-zero emissions towards setting carbon-negative targets, which would repay some of the carbon debt on its books from decades of unconstrained fossil fuel use.

In some ways it's shocking that the country that launched the coal-powered Industrial Revolution has become the global leader on cutting emissions. But what's happened in the UK over the last twenty years or so clearly shows how political tailwinds, sound analysis and smart legislation can help a country align with science-based climate goals. The climate fight is finally starting to use the law to its benefit, instead of the law being a drag on cutting emissions.

Worthington is an unlikely hero in the story. She studied English at Cambridge University and started her career in an environmental group with a focus on conservation and biodiversity. But in the mid 1990s she heard scientists raising the alarm that conservation work would not be enough to stop many species from going extinct because of global warming.

'Everything we care about is threatened by climate change,' she says. 'And I mean everything. Not just biodiversity, but also social stability.'

So she began to work on climate change full time, starting as a campaigner for the environmental charity Friends of the Earth (FOE). In the 1990s the UK was steadily cutting its emissions. The government celebrated and took credit for the progress, but a closer look at the numbers showed that the savings were not because the country was weaning itself off fossil fuels, but rather because it was switching from especially high-emissions coal use to burning less carbon-intensive natural gas. That was largely down to a recent discovery of a huge cache of gas in the UK's North Sea.

When the Labour Party came to power in 1997 with a crushing defeat of the Conservatives, it signed the UK up to the Kyoto Protocol under the United Nations framework. That committed the country to reducing its greenhouse gas emissions by about 13% by 2012, relative to 1990 levels. Because the coal-to-gas transition was already underway, that goal seemed easily achievable to the Labour leadership, and the party's election manifesto promised a 20% reduction goal instead.[2]

But as an ostensible champion of working-class people, the Labour prime minister, Tony Blair, also promised to save coal jobs, and, once in power, put an end to building new gas power plants across the country in 1997 – thus helping to extend the lives of coal power plants.[3] That caused emissions to stop falling. Most of the other laws Blair's party passed in the nineties only made incremental changes towards reducing emissions, which meant the ambitious 20% goal the party had promised in its manifesto began to slip away.

Worthington, a self-confessed data geek, realized that the UK government had no coherent strategy to cut emissions. The ministry in charge of environmental issues did not have the power to enact policies that would address the problem, and the ministry in charge of energy issues didn't much care about climate change.

In the early 2000s she worked with colleagues at Friends of the Earth on a report that called on the government to put a climate law in place that would align all ministries towards the same goal. The FOE called it the Big Ask Campaign and said the government should set timebound carbon budgets all the way to 2050 in line with what climate scientists were calling for.

'When you are on the outside and you are lobbying, you kind of hope that you will have some impact,' Worthington says. 'But you're never really very sure.'

While the Labour government showed some interest in the report, support mostly came from cross-party parliamentary backbenchers – that is, those without any official role in government or in opposition – who didn't really have the power to do much. Meanwhile, Worthington was hired by Scottish and Southern Energy (SSE), a power and gas company in Scotland and Northern Ireland. The company's boss understood the urgency of acting on climate change and invited Worthington to assess the realities on the ground. 'You can tell me how to run my company,' he told Worthington, 'and I'll tell you how to be a better campaigner.'

At the time, SSE was running its coal-fired power plants as hard as it could, whenever the price of electricity made it profitable. Worthington saw at first hand what she had witnessed in the UK's greenhouse gas inventory spreadsheets. As long as coal power made economic sense, no power company would turn it off.

She also realized that was where the opportunity lay. Electricity forms the backbone of the modern economy, and most countries heavily regulate utilities to ensure energy security. Worthington hypothesized that those regulations could be amended to make coal uneconomical at the pace needed to meet climate goals.

As luck would have it, an opportunity to do something about those regulations came just at the right time. In 2007 Worthington got a call from a former FOE campaigner who was now working as a civil servant in the UK government. Would Worthington consider coming back to London?

Between the FOE's Big Ask Campaign and Worthington's return to government a few years later, the Conservative Party had lost its third consecutive general election. To shake things up, the party in 2005 appointed an energetic new leader, David Cameron. The opinion polls told him that Brits now cared a lot about climate change – thanks in part to increased media coverage of the crisis. He wanted to use the issue as a way to 'decontaminate' the Conservative brand. He promised to create a 'carbon audit office' if he were to become prime minister at the next election, and to end the 'Westminster party dogfight' to bring about a consensus on climate change. Cameron even changed his party's logo from a hand holding a torch to a scribbled oak tree. The party slogan for the 2006 local elections in May was 'Vote Blue, Go Green'.

It worked. The Conservatives made gains over Labour, picking up more than 300 new elected officials across the country's councils. The victory helped Cameron go a step further. In September that year he promised to introduce a climate change bill in response to Friends of the Earth's campaign.*

The Labour Party's poor performance in local elections, combined with Cameron doubling down on a green agenda, proved to be a wake-up call. In one of the many steps Labour took to regain public confidence, it appointed rising star David Miliband as the environment minister. Then considered a possible future prime minister, Miliband was keen to make his mark.

Soon after taking up the job he realized, as Worthington and her FOE campaign had done just a few years previously, that the environment ministry wasn't as powerful as it needed to be to cut emissions fast. In response, he secured permission from then prime minister, Tony Blair, to create the Office of Climate Change, a cross-ministry team that would help overcome some of the infighting that slowed down negotiations and made it difficult to take on big problems. Internally, the office came to be known as a safe space where bureaucrats could engage in frank conversations about the limitations of various policies and feel empowered to work with different government departments to convince them to get onboard with real solutions.

All this coincided with the British economist Nicholas Stern's now famous report looking at the economics of climate change. As the planet warmed, he found, so the world risked losing 5% of its global gross domestic product and as much as 20% in the worst case.[4] Stern called climate change the greatest and widest-ranging market failure ever seen.

---

*Earlier, in July, the band Radiohead had been the lead act in the Big Ask Live concert in London. Both David Cameron and David Miliband were in the crowd.

Big businesses also seemed to finally be taking notice. The Confederation of British Industry, the country's most powerful business lobby, showed its support. Many, including oil giants like BP and Shell, wrote to Blair arguing why the 'transition to a low-carbon economy' would be good for the UK.

Political pressure from non-profits and business, and the reframing of the issue as economic rather than solely environmental – and, of course, rising competition from the opposition party – helped create the conditions for the Labour Party to push a step change in the UK's approach to climate change.

Across the Atlantic in 2005, California's governor, Arnold Schwarzenegger, passed an executive order that established the target of reducing greenhouse gas emissions by 80% by 2050, relative to 1990 levels. In 2006 the Californian legislature backed those goals and passed the Global Warming Solutions Act, legally binding the state government. Though California was a state in a country, its economy was not that much smaller than the UK's. A British team of Labour Party members and government bureaucrats travelled to California to learn how the state had pulled it off.

Meanwhile, Worthington began working in the newly formed Office of Climate Change, which Miliband had tasked to come up with realistic options for how government could reduce CO_2 emissions and adapt to climate change. Could they promise a bill that would create a legal mandate to act? The problem was that it would require parliamentary approval and initially Miliband wasn't sure there were enough votes to get it through. However, Cameron's promise in September 2007 to introduce similar legislation, if his party came to power, pushed Miliband over the line and the government proposed bringing a climate change bill to Parliament later that year.

The foreign ministry and environment ministry were in favour, arguing that a climate change bill would instantly make the country a global leader tackling a global issue. But the UK's Treasury, the government's finance ministry, worried that setting such ambitious climate goals could severely harm the country's global economic competitiveness. It asked the Office of Climate Change to add a conditionality clause that would force the UK government to stay the course for now – and to act only if other countries were doing so.

Similarly, both the energy ministry and industry representatives were worried that if the costs of meeting these goals were too high then emissions-intensive companies would migrate to other more carbon-friendly countries rather than work to cut emissions. There was also the chance of reputational risk that, if the UK failed to achieve its goals, pro-fossil fuel camps would characterize the country as an example of why other nations should not pursue aggressive climate action.

Beyond the infighting within departments, there were debates about the very structure of the bill. The legislation was to include the creation of carbon budgets, which would decrease as time went on and thus force the UK to cut emissions. Should the budgets be annual or in five-year chunks? The latter provided the sort of flexibility that governments seek in any policy. But if flexibility were offered, it had to come with a way to ensure compliance.

The proposal to create a Committee on Climate Change, an independent watchdog, could ensure that the government wouldn't become complacent. But those in the more conservative camp hammered the point that the committee couldn't be given too much power. For example, it couldn't have authority to actually create policies. The optics of giving the power of making laws to a body not elected by the people would be horrible.

Ultimately, says Worthington, what perhaps helped most to reconcile these differences was the fact that the draft bill had to be put together in mere months. The Office of Climate Change proved handy, with its ability to work across departments and make quick compromises to keep the drafting process moving without long delays. The expediency of the process and the push from Miliband meant that many of the most crucial elements of the draft, published in March 2007, did not change very much from their initial draft form of months earlier. The bill was built on four pillars: a greenhouse gas emissions goal, a pathway to reach that target using carbon budgets, a toolkit of policies to start the process, and the creation of an independent committee as part of a monitoring framework that would ensure compliance.[5]

It was introduced in the House of Lords, the upper chamber of Parliament, in November 2007. Initially, the goal it laid out was to cut emissions by 60% by 2050, relative to a 1990 baseline.[6] Unhappy environmental organizations, including the World Wildlife Fund UK, launched a campaign to make the target more ambitious. Those efforts turned out to be relatively successful, with the newest report from the IPCC, published in 2007, backing it.

The act finally passed in November 2008 with a goal of 80% reduction by 2050, and also giving government powers to introduce regulations to help reach those goals. Perhaps most importantly, it enabled the government to create an emissions trading system – which is exactly what happened after Brexit, in January 2021, forced the UK to leave the European Union's carbon market.

Well before then, in December 2008, the Committee on Climate Change (now renamed the Climate Change Committee) published its first report, laying out a precise five-year carbon budget for the UK government to meet. Every year since, the committee has published annual reports showing whether the government is on track and what it must do to meet its goals.

Since 2008 dozens of countries – including Sweden, France and New Zealand – have passed national-level climate acts, taking the UK's and adapted it to national needs. And it's not just rich countries. Climate laws also exist in Bangladesh, Bulgaria, Micronesia and the Philippines.[7]

Still, there aren't comprehensive national climate laws in most countries – or even in all the largest emitters. However, this does not mean that other laws or common legal practices cannot be used to hold governments accountable for the climate promises they have made.

In 2013 climate activists and the non-profit Urgenda sued the Dutch government for setting inadequate climate targets. The IPCC's fourth assessment report had found that rich countries needed to cut emissions by 25% by 2020 relative to 1990 levels to ensure that the world does not warm beyond 2 °C relative to pre-industrial levels. The activists argued that the Dutch government's climate promise must follow the science and the government had a duty to ensure that.[8]

In an initial 2015 ruling the Dutch district court that heard the case found the activists' argument reasonable and demanded the government step up its efforts to reach the 25% goal by 2020. The government appealed, and the activists in response argued that failure to meet the goal would be a dereliction of its legally mandated duty of care, because climate change would endanger citizens' fundamental right to life.

This duty-of-care argument has legal precedent going back decades. In 1932 May Donoghue and her friend went to a café in Scotland. Donoghue ordered a Scotsman ice-cream float – ginger beer poured over ice cream. The owner of the café brought the ice cream in a tumbler and poured the ginger beer from a brown opaque bottle. After Donoghue consumed some of her ice-cream float, her friend poured more ginger beer over – and a dead snail tumbled out of the bottle. That night Donoghue felt ill, and some days later had to be treated for shock and gastro-enteritis. She sued the ginger beer maker and won the case after a lengthy battle that went all the way to the highest court at the time, the House of Lords.

That set a legal precedent for a duty of care to the people being served. It's not just that every ginger beer maker must ensure that the product it sells does no harm to customers, but every entity selling products or providing a service must do so too.

Many decades later, in the mid 2010s, climate activists argued that if the Dutch government was not working rapidly to reduce greenhouse gas emissions then it was in violation of the same duty of care to protect the fundamental human right to live safely on this planet. The power in such an argument is that fundamental rights are sacrosanct in most countries in the world. In addition, the climate science underpinning the IPCC reports – the summaries of which are signed by nearly all nations after a line-by-line review – is strong evidence that inaction on climate change constitutes a neglect of duty of care.

The Dutch government's appeals went all the way to the country's Supreme Court, but to no avail. The climate activists and Urgenda won, and the court ordered the government to raise the ambition of its climate target.

These successes have emboldened activists to get creative and use the Urgenda case to go after private companies. In 2019 the Dutch arm of Friends of the Earth sued then Netherlands-headquartered Shell on the same duty-of-care violation. It alleged that Shell's climate targets were not in line with the Paris Agreement and the company must reduce emissions 45% by 2030 relative to 2010 levels. After two years of hearings, the group won in a lower court. Shell is currently appealing the decision.

A similar court challenge was brought in Germany in 2020, when activists supported by non-profits Germanwatch and Greenpeace sued the government for setting climate targets that didn't match up with the Paris Agreement. The group alleged that Germany not doing its fair share to cut emissions was a threat to the fundamental right to life and physical integrity enshrined in the country's basic law.[9]

Unlike the Dutch case, which dragged on for years, Germany's federal constitutional court ruled almost right away. And as a result, in 2021 the long-serving chancellor, Angela Merkel, committed to raising the country's climate goals. As one of her last major moves before leaving office, Merkel committed Germany to bring its net-zero goal forward from 2050 to 2045, which made it the most ambitious target among major economies.[10]

But the success of such cases isn't guaranteed. In Australia in 2020, eight young people used a duty-of-care argument to call for an injunction to stop the government from approving a new coal mine. The Federal Court of Australia accepted the argument as valid and created a precedent for duty of care, but nevertheless declined to provide an injunction. The government appealed against the duty-of-care finding, and a higher court agreed to overturn it. In its reasoning, it said that questions of policy are 'unsuitable for the judicial branch to resolve'. In other words, courts may be over-reaching in interpreting the duty of care as it connects to greenhouse gas emissions and venturing into territory that instead is the task of government and elected legislative bodies.[11]

The creative use of law towards climate outcomes is for some climate activists a tactic of last resort. Under the Paris Agreement each national government is supposed to reduce emissions towards the shared goal of keeping average global temperatures from rising indefinitely, but there is no set punishment if a country fails at its self-set goals.

'National courts offer a way of advancing some form of hard accountability with real sanctions attached to those climate commitments,' says Tessa Khan, founder of the Climate Litigation Network. In a little more than a decade nearly 2,000 cases have been filed around the world's courts to force governments and corporations to do more to act on climate change.[12]

Still, Khan is the first to acknowledge that court-won battles do not always lead to the changes needed. In the 2013 Dutch case, for example, the government dragged its feet, appealing repeatedly until the final judgment came at the end of 2019 – only a year before the 25% emissions reduction target. It took a pandemic-related slowdown in the economy for the government to get on track to meet that goal. Had that not happened, Khan thinks, the government could have got away with it by arguing that it had tried its best. That can be a valid defence in duty-of-care cases.

While losses in courts may shame a government, real accountability comes from the democratic process. The ideal situation would be a country where citizens elect leaders with a climate mandate, forcing those leaders to create climate laws with broad acceptance across the political spectrum.

It is already happening in some places, where demand for climate action is growing in popularity so quickly that not meeting climate targets can become political suicide. Australia, for example, saw six prime ministers in over a decade as it swung wildly on climate policies, with the most recent election in 2022 providing the clearest mandate to do more on climate.[13]

In places where the combination of factors has been in place for a while, progress can be fast. The UK has had six different prime ministers since 2007, when conversation around the Climate Change Act began. Despite coming from different political parties and with very different policy ideas, every single one has strengthened the country's climate targets.

Still, even in a place like the UK, climate activists may need to turn to the courts. The Climate Change Committee warned in 2021 that without adopting more ambitious policies soon, the UK may fail to cut emissions fast enough to meet its goals. As a result, non-profits Friends of the Earth, ClientEarth and the Good Law Project sued the government in 2022 for failing to set the necessary climate policies.[14]

The court case on its own may not be enough to compel the government to change. But it also enables the conversation around climate goals to happen in public and that forces the government to explain its actions. Climate laws mean governments not meeting their targets can be sued. However, the bigger impact comes from reputational damage. With demand for climate action growing in popularity, not meeting climate targets can become political suicide for some.

Democracy can be the biggest way to rein in the excesses of capitalism. While not all countries in the world are thriving democracies, it is the system that runs a majority of countries, including most of the largest emitters. If anything, as economist and journalist Martin Wolf argues, neither capitalism nor democracy can survive and thrive without the other.[15]

Still, even with a willing population giving politicians a climate mandate, it requires careful thought to ensure that climate laws work as intended in cutting emissions. The UK's climate law provides a framework that each country, with its unique needs, can tweak. Outside the realm of government, laws provide the market signal for corporations to chart out their future. As the biggest unit of capitalism, companies are learning to be a part of the net-zero world. If not, the biggest losers will be shareholders, who aren't going to sit by quietly as the tragedy unfolds.

I I

The Capitalist

The world's most sustainable company makes toilet cleaners, deodorants and mayonnaise. Unilever, with brands like Domestos, Lynx and Hellman's, has been ranked first every year since 2011 by GlobeScan, which surveys leading sustainability professionals.[1] And it is miles ahead of others that make the top of the list, including Tesla, Microsoft and Ørsted.

That's a remarkable achievement for a company with a huge variety of products, which Unilever says some 2.5 billion people in the world use every day.[2] Few corporations in the world have such reach. The story of how a consumer goods company has a higher sustainability rating than Elon Musk's Tesla, Bill Gates's Microsoft and the only large oil company to go fully renewable, Ørsted, is key to understanding how corporations can be a part of the climate solution.

Unilever seems to have found a way of turning corporate social responsibility plans from nice to have into the core of its business model. In a world of almost identical consumer goods differentiated only by packaging and branding, Unilever has made doing good a key selling point for consumers and attracting top talent.

If anything, the sustainability push may have saved Unilever. Between 1990 and 2010 the company was losing market share to competition from US giant Procter & Gamble (P & G) and Swiss company Nestlé. 'Unilever got terribly off track to be honest,' says Paul Polman who served as CEO of Unilever between 2009 and 2019. 'It was run by finance people for short-term profits, chasing targets that they couldn't deliver.'

The result was cutting spending on training employees and brand development just to prop up numbers that would please shareholders. 'It was a milking strategy,' says Polman, 'And you get into a downward spiral.' There was a risk Unilever wasn't thinking about the long term and milking itself so dry of resources it wouldn't recover.

Through past experiences at P & G and Nestlé, Polman understood that Unilever needed something different. Aged fifty-two at the time, it was likely going to be the last big job he'd take at a multinational and he wanted to use his power as the new CEO in 2009 to also do something bigger: reform capitalism. 'I wanted to make it clear that Unilever wasn't just working for the shareholders,' he says. 'We were optimizing for multiple stakeholders: the people who worked for Unilever, our customers and the planet.'

Such phrases about 'making the world a better place' are easy to find in investor brochures, but greenwashing is rife in the corporate world, and companies rarely live up to those promises. And Polman wasn't saying anything that many reformists hadn't already said before: running capitalism for short-term profits is going to mean less value in the long term. He wanted to show that a more sustainable form of capitalism is possible. The surprising thing is that, in his decade at Unilever's helm, Polman actually did it. The company's market capitalization grew threefold during his tenure; revenues rose 30%, and direct emissions more than halved.[3]

The complete opposite of Unilever in this regard is the oil giant ExxonMobil. The world's most valuable corporation as recently as 2013, it has experienced a decade of turmoil.[4] Sustained resistance to acting on climate change and years of poor financial performance finally led to a rebellion at the highest level in 2021.[5] That year, a tiny climate-activist hedge fund won enough votes to replace three directors on the company's board of twelve, who are ultimately responsible for the direction the company takes in the years to come.[6]

In their journey to be a part of the climate solution, most companies fall somewhere in between Unilever and Exxon – with the majority still closer to the problematic end of the spectrum. Lessons drawn from these two extreme examples can show what corporations – the engines of capitalism – will have to do, or what they may be forced to do as the planet heats up.

Although we differentiate between private and public companies, ultimately it's people that own corporations. Ownership structures can get complicated – with stakes held through trusts, shell companies and limited liability partnerships – but they always end with people. These are the capitalists running the system, and they take the lion's share of the blame for the climate mess. Between 1990 and 2015, an Oxfam study found, the world's wealthiest 1% were responsible for more than twice the emissions of the bottom 50%.[7]

Democratic institutions (at least where they exist) are supposed to provide representation regardless of a person's wealth, but the rich are known to distort those outcomes with sizeable political donations. The clearest and most direct way in which the wealthy wield influence, however, is through corporations.

Even in so-called public companies it is a relatively small number of actors that decides the direction. The largest shareholders in most companies are not individual billionaires but institutional investors like BlackRock and Vanguard. Though many of us with pensions own shares in companies like Shell, these stakes are typically controlled by institutional investors. In exchange for getting a decent return on the capital we invest, we give them our right to vote on the direction the company takes. Crucially, though, Shell is a multinational company and most of its owners are in the richest countries, which further distorts its representative capacity.

Given these problems, many environmentalists believe that the only long-term solution to tackling the climate crisis is to uproot capitalism and replace it with something that puts the right value on nature and human life. There is no way to ensure human well-being without also ensuring nature's well-being, they argue. If corporations are the source of so much that's wrong with the world, they say, perhaps we should try a government-led reshaping of the world.

It's tempting to believe that a benign dictator or technocratic management of the economy would be better for solving the climate problem. But the case is weak, with Russia, Saudi Arabia and Iran proving to be some of the biggest hurdles in taking global climate action.[8] Similarly, centralized economies, such as North Korea or the dissolved Soviet Union, have shown disastrous failings too.[9]

Crucially, capitalism, from the United States to China, is now deeply entrenched. There's not one 'utopian' country in the world where it has been successfully replaced by an economic system that solves climate change.[10] Even if you think the best option is to overthrow capitalism, there doesn't seem to be any way that a new system could be put in place within the few decades left to avert catastrophic climate change.

What may be possible within that tight time frame, however, is reforming capitalism. But how?

One of the first things Polman did when he became CEO of Unilever was suspend reporting on quarterly profit guidance.[11] The figure is often used by analysts to advise clients on whether the company's stock is worth holding, buying or selling. It can increase volatility in the market and become a headache for CEOs.

'Business has become short term; it reacts to the symptoms not the causes,' says Polman. 'So I thought, if we really want to make this business work again, I need a little more time.' Without the pressure of quarterly profit guidance, Polman was able to change the rhythm of the investments Unilever could make. That was necessary to undo the under-investment of his predecessors, which he diagnosed to be one of the problems behind Unilever's falling performance as a company.

Apart from tweaking the financial flows, Polman also needed to reinvigorate long-term thinking among Unilever's employees. The first big meeting of executives that he convened after becoming CEO was in Port Sunlight near Liverpool in the UK.[12] He wanted to remind the company of the value of its founders, who had built the port as a model employee village. Founded in 1885, Lever Brothers, as it was then called, offered staff shorter work days, health benefits and savings plans – perks that today we see as the bare minimum companies ought to offer, but which were far ahead of that offered elsewhere back then.[13]

Polman changed over 70 of the 100 top management people of the company, which employs nearly 150,000.[14] He reintroduced training programmes, adjusted compensation plans that would invest short-term bonuses into long-term savings, and added a slew of employee benefits in support of physical fitness, mental health and flexible working. 'Ultimately, the company is an aggregation of people,' he says. 'You cannot have a sustainable business model if the people themselves aren't living sustainable lives.'

He also began a review of Unilever's many different sustainability initiatives, which were then mainly focused on fisheries, agriculture and water. Thomas Lingard, now global sustainability director, was on the sustainability team at the time and was tasked with adding climate to the portfolio. 'Before Paul, sustainability work was in the "responsible business" bucket,' he says. 'Paul was really, really driven to put it at the centre commercially.'

The business case for sustainability had three pillars, according to Lingard. First, customer expectations were changing and, he argued, they would soon care more and more about buying products that were not harmful to the planet. Second, many of the steps to becoming more sustainable, such as becoming more energy efficient or investing in renewables, would save the company money in the long term. And, finally, if the company became part of tackling climate change then it would reduce the risk to its business, which was dependent on sales in almost every country on the planet.

Climate impacts would eventually make the world head in the direction of becoming more sustainable anyway, says Lingard, and because consumers' demands change so rapidly, 'being slightly ahead of trends is good in our industry'.

Take palm oil. You may have never heard of it, but there's zero chance that you have not bought a product that contains it. Palm oil, a type of vegetable oil, is used in everything from toothpaste to lipstick and dog food to biofuel, and global demand for the commodity is projected to increase by 46% by 2050.[15] But there is a problem.

In 2020 a study found that more than 90% of palm oil comes from Borneo, Sumatra and the Malay Peninsula in South East Asia, which are home to some of the world's most important rainforests. About 50% of deforestation in Malaysian Borneo is directly linked to the production of palm oil.

Despite this, if the world has to use some form of vegetable oil in consumer goods, palm oil is perhaps the best bet today, because it has a very high yield to a relatively low land use footprint. Of the total global demand for vegetable oil, palm oil makes up 40% – while accounting for about 5% of the land dedicated to vegetable oil production. That means the world needs to find a sustainable way to produce palm oil instead, and Unilever is one among a handful of companies that have been working on a solution since the 2000s.[16]

Unilever was one of the earliest backers of the Roundtable on Sustainable Palm Oil, which has created a certification system that aims to ensure the product isn't linked to deforestation. It does that through understanding the complexity of palm oil supply chains and better tracing the source of the commodity. About a fifth of the world's palm oil supply is now considered sustainable.[17]

'Eventually people will understand what's going on with forests and there will be a backlash against unsustainable palm oil,' says Lingard. 'But you can't fix these things overnight. It takes years to build a supply chain.'

That's one reason why, under Polman's orders, the team created the Unilever Sustainable Living Plan, which set a goal to decouple the company's growth from environmental impact. Specifically, Unilever would halve the environmental footprint of making and using its products, even if the volume of products it made increased.[18] It was by far the most ambitious sustainability goal set by any large company in 2010.

While the sustainability push was underway, Polman's biggest test came in the form of one of the rudest manifestations of capitalism. In 2017 Unilever received an unsolicited takeover bid from Kraft Heinz, which was owned by private equity firm 3G Capital and Warren Buffett's Berkshire Hathaway.[19] The offer was to buy Unilever for $143 billion, 18% more than the company's market value at that point.[20] The company's stock price soared when news of the bid leaked.

The company was seven years into Polman's plan and was seeing positive results: it had regained market share, grown in valuation and reduced its environmental footprint. But Kraft Heinz's bid sent a message that the company was leaving value on the table. Or as *Bloomberg Businessweek* explained, 'Your CEO is spending too much time keynoting climate-change conferences, when he should be finding ways to move more mayo.'[21]

For two companies that made similar products, their business models could not have been more different. On one end of the spectrum were companies like Kraft Heinz, which were focused on cutting costs, increasing margins and paying little in taxes. The *Financial Times* described the model as one 'that ultimately destroys businesses by starving them of investment'.[22] On the other end was Unilever, sacrificing some profit today and caring for the environment with the theory that this would ultimately be better for shareholders in the long term.

For a takeover bid to succeed, the majority of a company's board of directors need to vote in favour. But the same board that had approved Polman's sustainability plan wasn't interested in having those changes rolled back by Kraft Heinz, whose bid was masterminded by 3G Capital. In Polman's view, 3G was the perfect example of what happens at companies beholden to 'shareholder primacy': cutting costs and juicing profits.

When a board is reluctant to hand over the company to a bidder and the bidder persists, the takeover becomes hostile. The bidder then has to convince the company's shareholders to vote enough new directors to the board who favour the takeover to overcome opposition to the bid. 3G had never failed at a takeover bid; according to *Businessweek*, it had 'a history of grinding down opposition'. When it went after the brewer SABMiller, it raised its bid four times and piled so much pressure on the company's board that its directors could no longer refuse to let investors reap the windfall.

The deal would have made Polman a lot of money personally. 'The guys from Kraft Heinz told me, "Paul, you should sell out to me. On top of that you can get $200 million and you'd be very rich",' he says. 'I said, "Great, give me $200 million and I will spend all of that against you."'

Over a period of nine days, 3G would be handed its first defeat. In Polman's telling, the work he had begun with the sustainability plan is what helped him defend Unilever against 3G. John Sauven, former head of Greenpeace UK, which in 2008 sent activists to climb up Unilever's headquarters in protest, came to the company's defence.[23] So did Ron Oswald, general secretary of the

International Union of Food Workers, which represents 10 million workers in agriculture and hospitality, who said: 'Kraft Heinz was the epitome of what a company should not be: pure financial engineering.'

After Unilever made a public statement that there was 'no merit, either financial or strategic' in a deal, *Businessweek* reported Polman pulling his strings, too. He fired off a letter to the founders of 3G and to Buffett making the same points. His staff started telling journalists that the company's long-term business model would be destroyed by the takeover.

In the hostile bid, Kraft Heinz needed to turn shareholder sentiment in its favour. That meant launching a public relations blitz, for which it hired UK-based elite PR agency Finsbury. But Finsbury's majority owner was advertising giant WPP, which handles much of Unilever's multibillion-dollar marketing budget. Polman wrote to Martin Sorrell, WPP's CEO, and reportedly within hours Finsbury dropped the new client.[24]

As the fate of one of the UK's largest companies hung in the balance, even the country's prime minister, Theresa May, began to worry. Would the new owners cut jobs? Would they eventually move the company outside the UK? It would be a hit to the country's industrial base, which had been in decline for decades. Not a concern to have at a time when May was trying to steer the country towards a Brexit deal with the European Union.

Realizing that the odds were stacked against it, 3G dropped the bid. Polman breathed a sigh of relief. He believes that his success in standing against the takeover is down to Unilever proving its sustainability-linked business case. 'Increasingly, consumers want to identify themselves with the products they buy, if it mirrors their lifestyles, if it mirrors their philosophy, if it mirrors what they think needs to change,' he says.

Even with a market valuation of $130 billion as of April 2023, however, Unilever is a tiny cog in the capitalist engine. In the past decade more and more companies have begun taking sustainability seriously, but the pace of change is too slow to address the climate crisis. 'There's a little bit of momentum,' Polman says. 'The problem is no longer the direction we are headed in. The problem now is speed and scale.'

Corporations have typically been a barrier to government policies aimed at cutting emissions. And as big businesses have grown their share of global economic output so their grip on governments and regulations has grown. Companies are easily able to outspend almost anyone in lobbying politicians, and this pushes political will against regulations that could lower profit. Capitalism works better through competition. But when certain entities within the system get too powerful, they can squeeze the system to their benefit. This kind of crony capitalism shows why the libertarian ideals of free markets untethered by government regulations tend to fail.

That's why, after finishing his stint at Unilever, in 2019 Polman launched Imagine, a non-profit, to bring CEOs together to agree on common goals. Instead of competing against each other in a race to the bottom to juice short-term profits, he hopes Imagine can create a coalition of CEOs to collectively act on environmental and social issues. So far, he has forged alliances among companies in food, fashion and finance, all of which are committed to specific climate goals.[25]

'Everybody complains, but we work with CEOs until everybody compromises,' he says. 'Ultimately, we de-risk government action.' That's because once a significant portion of the industry aligns with certain goals, say, a policy to phase out coal or agreeing to disclose climate plans to reach net-zero emissions by 2050, then the group can expend its lobbying power in unison towards it.

If any company attempts to go green alone – a process that inevitably involves upfront costs – it can lose the edge to a competitor that does not follow the path. That's why, even if addressing climate change is likely to prove economically beneficial in the long run, the short-term hit can be a serious barrier. Efforts like Polman's Imagine create the room for building non-competitive coalitions, where companies all commit to the same green goals and thus lower the risk of losing out.

But these voluntary commitments as part of coalitions are not the only way to act. A firmer way would be if a majority of shareholders demanded a company do right by the environment because they believe the long-term financial performance of the company would be better for it. In December 2020 a small hedge fund called Engine No. 1 launched a campaign to appoint new directors to ExxonMobil's board and get the company to change its direction from doubling down on fossil fuels to working on a plan fit for the energy transition. It's a David vs Goliath battle with implications for whether capitalism is a solution for a planet on fire.

Before we get to how the fight played out, let's remind ourselves of what unhappy shareholders can do to make a company like Exxon change its course.[26]

They can put out a press release or tell journalists why Exxon continuing to rely only on fossil fuels and not investing in renewables puts the company's future at risk. Company management, especially CEOs, prefer to be liked and respected. A bad news story can be embarrassing, but does not force a company to do anything.

A shareholder can ask others to vote on non-binding resolutions that would ask Exxon to, say, prepare a climate impact report. If a majority vote in favour then the company is under more pressure than usual to do something and it might just produce the report, as Exxon did in 2018, saying that the company's strategy posed no material risk to its business. (Experts labelled the report 'defective and unresponsive'.)[27]

Each year, Exxon's management has to ask shareholders to approve executive pay packages. If the majority vote against, that also adds pressure although the vote remains non-binding and the senior executives at Exxon can get their compensation anyway.

Similarly, each year Exxon's board of directors are up for election, and shareholders can vote against them. If a majority vote against a director then the person may be required to hand in a resignation and Exxon will have to nominate someone else. Still, the board does not strictly have to accept the resignation.

Of course, if an Exxon shareholder is unhappy with what Exxon does in response to these votes, they can always sell their company shares; indeed some climate activists have successfully pressured large shareholders, for example university pension funds, to divest from their recalcitrant fossil fuel companies. In 2020 the Church of England Pensions Board did exactly that. After Exxon repeatedly failed to set goals to reduce emissions from customers' use of its fossil fuels, the Church's board sold its stake in the company.[28] The theory goes that if enough shareholders divested from Exxon then its stock price will go down, which will hurt the company's ability to raise money and perhaps eventually take away its social licence to operate.

However, a divestment strategy does not do much to hurt a company's fundamental business. If a company is making heaps of profit extracting and selling fossil fuels, divestment from a green-minded shareholder just makes it easier for someone else to benefit even more. Those profits also ensure that access to capital remains easy for the company. It's one reason Bill Gates held out on divesting from fossil fuels, despite a years-long campaign by activists.[29] In 2019 he finally did it, saying that he did not want to be profiting from the destruction of the planet. But he warned activists that it wasn't going to do very much to change the trajectories the companies take.

There are two other strategies shareholders can take to enforce a change. One is the above-mentioned hostile takeover; another is a proxy fight lodged through efforts to take control of the board.

In a hostile takeover a large fund could convince other shareholders that it's time to take the public company private. In return, the public company's shareholders would typically be rewarded with a share price that's higher than what the stock market offers. Once private, the new owners can fire the board and the CEO and replace them with those that agree with a greener strategy.

But a hostile takeover becomes harder the bigger the company. Kraft Heinz failed to do that to Unilever at a market capitalization of $143 billion, despite offering an 18% premium on the stock price. In December 2020 Exxon's valuation stood at about $160 billion and so there was little risk of a takeover.

That's why Engine No. 1 pursued a proxy fight. This is where a shareholder can write in names of new candidates for the company's board, arguing that they will steer the company towards a direction that's better for the company and thus for all shareholders. Specifically, Engine No. 1 argued that none of the directors on Exxon's board had any expertise in the energy industry. It is 'just common sense that an energy company should have at least some people with energy experience on the board', Chris James, founder of Engine No. 1, told the podcast *Capitalisn't*.[30] That they didn't, he said, 'tells you a lot about the kind of culture of the company. They didn't really want to be questioned.'

James was also hoping to harness Exxon shareholders' frustration with the company's repeated refusal to accept the reality of the current energy transition, to disclose all its emissions or to set out a science-aligned strategy to reduce emissions that would help the world meet the Paris Agreement goals. But 'we didn't talk about climate change as an ideological issue', he said. 'We spoke about it consistently as an economic issue.'

A proxy fight is not cheap. The shareholder going on the offensive has to first spend money finding and convincing the right candidates for board seats, and then has to spend much more selling those candidates to all the other shareholders, whose backing is crucial. When Engine No. 1 launched its campaign, it had a budget of $30 million for the Exxon battle.

That kind of spend can make a proxy fight a freerider problem. Engine No. 1 was betting that if it succeeded then it would lead to an increase in the Exxon share price, which could cover the expense of the proxy fight. In that case, all the other Exxon shareholders would see the value of their portfolio go up, even though they weren't spending any of their own money. But if Engine No. 1 lost the fight then it alone would bear all the cost, the other shareholders don't lose anything. Thus, there's little incentive for other shareholders to join Engine No. 1's campaign and more incentive to just watch what happens.

When the proxy campaign began, in December 2020, Engine No. 1's Exxon stake was a mere 0.02% or worth about $30 million.[31] The hedge fund was ready to spend almost the same amount of money in cash on the proxy campaign. That means Engine No. 1 was betting that it would win the board seats in the vote scheduled for May 2021, and Exxon's stock price would double – at least.

James's calculations showed that, if they could convince three of the biggest shareholders of Exxon then they had an 85% chance of winning enough votes to put at least some of its list of candidates on the company's board. Those three – namely Vanguard, BlackRock and State Street – owned about 20% of the company between them.[32]

The Big Three are institutional investors that manage money on behalf of clients – many of whom are ordinary, middle-class people with just enough to invest a little here and there.[33] Indeed, there's a good chance that some of your pension investment is held in one of the funds managed by them. They gained this status after the financial crisis of 2008–9 when the index fund industry boom began.

A stock index aggregates the changes in stock price of a defined set of companies. Which companies make it to an index can vary depending on the criteria set out by the index. The S&P 500, for instance, has 500 companies that are meant to 'represent leading companies in leading industries'.[34] Once created, however, an index can be very powerful. When the US president says that the stock market is booming, he is not basing the comment on one company or one sector but on something like the S&P 500. Popular indexes such as the UK's FTSE 100 or Stoxx Europe 600 can become indicators of the health of the regional economy. You're likely to see the changes to those indexes plastered on the front pages of major news publications.

From a fund manager's perspective, an index is a safer bet than a single stock. Because it consists of a broad set of companies in a big list of sectors, there's little risk of losing money just because one company or industry has a big problem. Crucially, years of financial analysis have shown that financial returns from investing in popular stock indexes can often outperform those who bet on specific stocks.

Most importantly, because an index fund manager doesn't have to do the work of picking what stocks go in a fund, they can charge lower fees for managing your investments. For example, a typical index fund would charge about 0.1% in annual fees for the total sum invested. On the other hand, managed funds can charge as much as 0.5%. The lower the fees the higher the share of the returns an investor can keep.

The combination of lower fees, lower risk and pretty decent returns has made index funds very popular. That's helped the Big Three capitalize on the trend and why they end up owning large portions of most public companies in the world – including often more than 20% of US companies.[35]

That gives the Big Three significant voting power at many of these powerful companies, which until recently they often wielded to reject climate resolutions.[36] That has led to climate activists targeting BlackRock, Vanguard and State Street for abandoning their duties to safeguard long-term return for investors and thus incorporate in their decisions the financial risk that climate change poses.

The activists' ire helped make Engine No. 1's case stronger. The investment firm needed the Big Three's support for the Exxon proxy vote and the Big Three were under pressure to show that they cared about what their companies did on climate change.

Meanwhile, Engine No. 1 continued to pressure Exxon directly. A private conversation between Exxon CEO Darren Woods and Engine No. 1's chief investment officer Charlie Penner took place in January 2021. It was reported by the *New York Times* as follows.[37]

Woods talked about how the company would play an important role in meeting the energy demands of a growing global population with improving standards of living. He said ExxonMobil supported the idea of addressing climate change but didn't know what kind of competitive advantage the company could have in areas like renewable energy.

'A lot of your investors think it would make sense to set longer-term goals,' Penner said midway through the call.

'Hey, Charlie, do you know how anybody is going to meet the 2050 goal today?' Woods replied. 'Have you asked any CEOs who have committed to that?'

'Do you know how you're going to fulfil your business plan without burning down the planet?' Penner asked.

'If all it takes is aspiration,' Woods said, and then paused. 'We support that ambition.'

'Have you ever accomplished anything that, when you started, you didn't know how you were going to finish?' Penner replied. To Penner, having a goal of getting to net zero even without an exact map was better business than planning to continue producing oil and gas in a decarbonizing world.

The call ended with the executives and activists saying they would continue to seek a resolution to avoid a standoff. They didn't speak again.

A few weeks after the call, Exxon added two new board members, one with experience in the energy industry and another in climate-focused investing, respectively.[38] Those additions expanded the board to twelve, and were a clear response to one of Engine No. 1's biggest concerns about Exxon's board: that it lacked experience in the energy industry and in the climate-driven energy transition.

That should have ended Engine No. 1's campaign. Exxon had listened to the criticism and done something about it. Most of the major investors are typically conservative in their outlook and tend to vote with the company's management rather than with activist investors. But Exxon's record both on financial performance and shareholder engagement had been so poor for so long that it seems that many shareholders felt Exxon hadn't earned the right to appoint new directors on its own.

So three months after the launch of the campaign, Engine No. 1's chances of winning didn't look too bad. Exxon wasn't going to sit and take it, though. If Engine No. 1 was looking to woo shareholders, the oil company was going to spend even more to counter the threat. It budgeted to spend $35 million, launching a new website, advertising on social media websites and throwing specialists at convincing shareholders to vote against Engine No. 1's board candidates.[39]

Convincing the Big Three to vote in its favour would net Engine No. 1 about 20% of the votes. It needed another 30% or more to ensure that the proxy fight would go its way. That's where so-called proxy advisory firms come in. Though they don't have a vote themselves, the advice of influential firms like Institutional Shareholder Services (ISS) and Glass Lewis can swing a further 10% or 20% of a shareholder base made up of smaller institutional investors.

Institutional investors hold stocks in thousands of companies, each of which might have many resolutions or board seats up for voting each year. That means these firms don't always have the time to consider each proposal on its merits, and they rely on the advisory firms to recommend which way to vote based on investors' preferences.

That means, if Engine No. 1 were able to convince proxy advisory firms and the Big Three, it was guaranteed a win regardless of Exxon's shenanigans. It got help from a forward-looking institutional investor to do just that.

Leading the charge was Aeisha Mastagni, a portfolio manager at California State Teachers' Retirement System, commonly known as CalSTRS. The US's second-largest pension fund, CalSTRS had created a name for itself by playing an active role in shaping the companies it invests in. Over the previous few years Mastagni and Engine No. 1's Penner, who was an activist investor with JANA Partners previously, had successfully lobbied Apple to add parental controls to curb addiction to devices among children, and got McDonald's to add plant-based burgers to its menu.[40]

When Mastagni brought the Exxon proxy fight idea to her boss Christopher Ailman, he was shocked. 'Holy bananas, can't we start with a smaller company first?' he recalled, talking to *Bloomberg* in June 2021.[41] 'Exxon is a behemoth and can be a bully.' But Mastagni convinced him to sign on; CalSTRS would work behind the scenes, drumming up support, while Penner and Engine No. 1 would be the face of the proxy battle.

That partnership proved essential, giving Engine No. 1 instant credibility when the campaign launched in December. Nobody at the time knew what Engine No. 1 was, whereas CalSTRS managed some $300 billion and owned about 0.2% of Exxon – which might sound small but accounted for ten times as many shares as Engine No. 1 had.

Then, when Exxon announced its own new board seats in January 2021 while it still did not have the climate or energy transition expertise that the activist investors were seeking, Mastagni began to organize webinars for proxy advisory firms and other large investors. She showed those investors how voting for their list of new board candidates was crucial for shareholders who care about Exxon's future on an overheated planet. As a result, both ISS and Glass Lewis supported Engine No. 1's list of candidates. 'I don't think it would have happened without her,' Penner told *Bloomberg*. 'She was incredibly forceful in her advocacy.'

While Penner and Mastagni made private calls to other Exxon shareholders, Engine No. 1's public spat with Exxon continued. Apart from appointing its own new board directors, Exxon also announced that it would spend $3 billion over the next five years on carbon capture (see Chapter 8) and other carbon reduction solutions. And the company continued to insist that its climate plan was already aligned with the Paris Agreement.

For a company that had resisted any calls for change for decades, these moves showed it was clearly rattled by the threat posed by a tiny hedge fund. 'What's maybe most surprising is that Exxon actually acted off the back of Engine No. 1's first letter,' said David Doherty, an oil specialist with BloombergNEF.[42]

Engine No. 1 asked David Victor, professor of innovation and public policy at the University of California, San Diego to take a deeper look at Exxon's claims. He read all the fine print on Exxon's many climate change presentations and published his findings in a white paper. It was not only an indictment of the entire oil industry – it showed Exxon was the laggard among the laggards.[43] 'ExxonMobil painted a future unaware of how the world of policy was changing,' he wrote. 'What remains is a shrinking group of oil majors, notably ExxonMobil, that still cling to old forecasting methods and results.'[44]

Separately, Engine No. 1 published its own analysis with insights from Victor's work, finding that Exxon's public statements painted a far rosier – or greenwashed – picture of where the company was on its climate change efforts than the numbers from the firm's carbon accounts suggested was the case.[45]

Based on the Greenhouse Gas (GHG) Protocol, a widely used but voluntary industry standard, companies report emissions in three categories. Scope 1 covers direct emissions, such as burning natural gas in a boiler to heat the office building. Scope 2 covers emissions that an electricity provider generates on the company's behalf to help it power its operations. And Scope 3 covers all the ways in which the company's supply chain and the use of its products lead to emissions.

What made matters worse for Exxon was that I, along with my *Bloomberg News* colleague Kevin Crowley, had seen documents in October 2020 that showed Exxon's pre-pandemic plan was to increase its Scope 1 and 2 emissions by as much as 17% by 2025. At the time the company did not report its Scope 3 emissions, but we estimated that they would have increased by a similar amount. In effect, the company's internal plans were going in the opposite direction of what the Paris Agreement needed, which is reducing emissions and fast.[46]

For oil companies, Scope 3 emissions are typically more than 80% of their totals because they cover the emissions produced when customers burn the fossil fuels they produce and sell. Though the GHG Protocol was developed in the late 1990s, its use became much more common after the Paris Agreement was signed. During the 2000s oil companies shied away from reporting Scope 3 emissions altogether, arguing that they did not have any control over how the customers used their products.

But, of course, that's ridiculous. The majority of the oil and gas they sell will be burned for fuel. And to ignore that fact is essentially to wish away the actual emissions impact of the industry. 'The future of the whole industry depends on what happens with Scope 3 emissions,' said Victor. 'So while they may be outside the control of any individual company, if the company has no vision for what they're going to do in a world where Scope 1, 2 and 3 must come down, then they have no vision on the liabilities of the industry that will be transformed or disappear.'

After years of pressure from activists, in the 2010s oil companies slowly began to report Scope 3 emissions and even take responsibility for them. Exxon was one of the last of the major oil companies to hold out – until the Engine No. 1 campaign. And even after the company reported the figure for the first time, in 2021, it appears to have offered no way to reduce those emissions and continued to deny responsibility for them.

The phone calls and letters proved to be enough. With the arrival of voting day, Engine No. 1's candidates had received the support of the Big Three and that of proxy advisors ISS and Glass Lewis. James and Penner were feeling good about their candidates' chances. But unlike voting in democratic elections, shareholders at a proxy vote can recall their vote and send in another with different choices right up until the last minute. Penner took the one last chance he got on 26 May 2021 to speak directly to the company's management and all the shareholders during Exxon's annual general meeting, before the voting deadline.

'Rather than being open to the idea of adding qualified energy experience to its board, we believe ExxonMobil once again closed ranks despite the increasingly diminished long-term returns of this approach. The good news, we believe, is that no matter what the outcome of today's vote, change is coming,' he said. Investors had come to an acceptance that driving humanity off a cliff is not good business, he added.[47]

Then something out of the ordinary happened. Exxon called for an hour-long recess to allow more time for the votes to come in. Shareholders couldn't recall if the company had ever suspended its annual meeting before. The Engine No. 1 team suspected that Exxon was using the time to influence votes, and their suspicions were confirmed when they heard from other shareholders who were getting calls from Exxon. To counter the narrative, Engine No. 1's PR firm started calling TV producers to get Penner on as a guest.

'It has a very banana-republic feel . . . We're aware that directors of the company are right now calling large shareholders and trying to get them to switch their votes,' he told CNBC while waiting for the recess to come to an end.[48] 'This is a classic kind of skulduggery and this is not the way to move this company forward.'

Exxon said that it was keeping the polls open because votes were still coming in. 'Is there a downside to giving shareholders more time to vote?' asked CNBC's Leslie Picker. 'Do you think that ultimately that would lead to potentially a more holistic result?'

'They are calling people who have already voted, and asking them to change their vote,' Penner replied. 'And they have the sole ability to close the polls as soon as they get enough people to say "yes". This is not a democratic expansion. This is the opposite.'

It didn't work out for Exxon. At the end of the recess the company said two of the four Engine No. 1 candidates had secured enough votes to gain a seat at the board, and one race was too close to call. It was a 'historic loss', concluded *Fortune*.[49] The CEO was dealt a 'stinging setback', said *Bloomberg*.[50] It was 'a day of reckoning' for Big Oil, noted *Vox*.[51] 'Wall Street rebels against Exxon', wrote the *New York Times*.[52]

A week later, on 2 June, Exxon said the race that had been too close to call had gone in favour of an Engine No. 1 candidate, giving them three seats in total on the board.[53] The company's stock price jumped to $65 – nearly double what it was when the proxy campaign was announced in December. The activist fund had budgeted to spend $30 million, but the final tally was only $12.5 million.[54] The increase in stock price more than paid for those costs.

Exxon's stock price has since doubled again, reaching $115 in April 2023. However, most of the recent upside is because of high oil and gas prices that followed Russia's attack on Ukraine. The company has used its record profits to defend its strategy of doubling down on fossil fuels, but that hasn't meant stepping back from climate-oriented investments. After the Inflation Reduction Act was passed in 2022, Exxon has also increased its ambition on carbon capture projects and even organized a low-carbon solutions spotlight day for investors wanting to know more about its strategy for the energy transition.

Winning those three seats was without doubt a corporate coup, but they still accounted for only 25% of the board, and Engine No. 1 wasn't under any illusion that change would happen right away. James was clear with shareholders voting in favour of his candidates that it would take Exxon as long as ten years to adapt their business model to see a world moving away from fossil fuels.[55] The victory, however, was proof that shareholders are capable of forcing a company's hand, regardless of its size or prestige.

Engine No. 1's success led to it launching an exchange-traded fund with the ticker VOTE. Those investing would get the returns of the S&P 500, but the hedge fund would use its voting power for environmental, social and governance (ESG) factors aligned with long-term returns. That is, the fund is no different from those you can get from the Big Three, except that you are told from the outset that the voting power will be used to support better ESG outcomes. The group also promised to continue shareholder activism, targeting companies that were severely lagging behind on climate.

When climate activists blame capitalism as the root of the climate crisis, it is likely that much of that ire is intended for the American economist Milton Friedman. In an essay written in 1970 for the *New York Times* magazine, he declared, 'The social responsibility of business is to increase profits.'[56] In nearly 3,000 words he laid out a vision of how the economy should work that has deeply divided opinion about the core function of business ever since.

A business is created to serve a specific purpose, Friedman wrote, such as producing goods or delivering services. Its executives are hired to do their best at steering the business towards that purpose and, in the process, create prosperity for shareholders. However, if the company is also asked to take on social responsibilities then it will distract itself choosing which causes to support and, worse, spend on those causes money that could have been spent on the business or returned to shareholders as profits.

More important for Friedman's vision of how a business should function, the act of choosing which social cause to spend money on is a political one, he argued – it was something an unelected corporate executive should not be dabbling in. If shareholders want to do social good then they may do so as individuals by using the profits their corporations produce.

There are major problems with his work, and they show why the root of the climate crisis is not capitalism but the corruption of capitalism.[57]

Marianne Bertrand, an economist at Friedman's alma mater, the University of Chicago, says that his essay relies on the naive belief that what's good for shareholders is good for society. And she believes that Friedman was probably aware of some of the flaws, which is why he wrote that companies should 'make as much money as possible while conforming to the basic rules of the society'. The implication was that governments would write the laws that realign profit maximization and social welfare. But too many lawmakers have essentially become the employees of shareholders rather than representatives of the people, she concludes, through the large donations they receive as political contributions from businesses.

Anand Giridharadas, author of the 2018 book *Winners Take All*, agrees with Friedman's belief that businesses should not stray from their lane – that is, making money – and should not assume government functions tending towards public welfare. But if that is the case, Giridharadas says, then Friedman is inconsistent when he supports corporations' goodwill expenditures 'that are entirely justified in its own self-interest'. In other words, companies should let governments create rules that allow private enterprises to create maximum good for all. But instead Friedman gives companies the moral cover to not worry about the public good and meddle in public affairs to rewrite rules for profit maximization.

Luigi Zingales, also an economist at the University of Chicago, says let's take Friedman at his word that maximizing profits for the shareholder should be the priority.[58] If the social goal is to support education then it doesn't matter whether a shareholder donates $1 to the local university, or the corporation does it. If the goal is to reduce pollution, however, the cost of stopping the pollution before it happens at the corporation is much lower than the cost of the shareholder cleaning up afterwards.

In living up to Friedman's doctrine of maximizing profits, businesses have cut costs and investment, even if that means employing people in precarious jobs, not paying them living wages, polluting the environment, avoiding taxes and influencing regulations to ensure none of those social ills violate any laws. That clearly is not a sustainable model for people or the planet. That's why you're seeing businesses starting to leave the doctrine behind.

In 2019, almost fifty years after Friedman's essay, the Business Roundtable published an open letter rebuking its central tenets.[59] It admitted that since 1997 it had 'endorsed the principles of shareholder primacy – that corporations exist principally to serve shareholders'. The new letter, signed by hundreds of the world's most powerful CEOs, redefines the purpose of a corporation. It says companies have a 'fundamental commitment to all of our stakeholders' and lists them in order: customers, employees, suppliers, communities we work in, the environment, and shareholders.

Polman's work at Unilever shows that businesses can do that without sacrificing returns to shareholders. And Exxon's loss to Engine No. 1 is proof that investors are willing to discipline companies that don't see the long-term future. It's no longer climate against capitalism; it's clear that increasingly the champions of capitalism want climate to be a problem that capitalism can solve rather than worsen.

12

The Next Steps

Getting to zero emissions on a deadline will mean changing everything. That sounds like an impossible task, but, as we've seen, the economic case is quite reasonable. We are constantly building and rebuilding, and it's possible to create a zero-emissions world within decades. And it is now certainly cheaper to save the world than destroy it.

It will be an enormous task. But given the transition has begun at scale, from China to California, we can learn from others and make it that bit easier. In every instance of scaling a climate solution, there is a combination of people, policy, finance and technology. However, the mix is different from country to country.

What works for solar power in India may not work in the US but is likely to work in Nigeria. What works for electric cars in China may not work in Europe but is likely to work in Saudi Arabia. What works for wind power in Denmark may not work in Chile but can work in Ireland.

As much as technological progress has been the brightest spot of the climate fight so far, it won't be possible to keep the good news going without creating a framework that helps the deployment of those technologies at scale. That may happen because of good laws (Chapter 10), supportive international institutions (Chapter 5) or accessible private capital (Chapter 6). If nothing else works, it's clear that capitalism's most powerful force – shareholders – are now ready to take matters into their own hands (Chapter 11).

How it will play out over the next few decades depends on us. It's just as important to understand the potential and limits of technologies as it is to understand the nuances of good policies and enabling entrepreneurialism.

This book has discussed many of the biggest climate solutions – solar, wind, batteries, electric cars, laws and policies – but it cannot discuss them all. For example, I could not say enough about hydrogen, lab-grown meat or the new ways to cut methane emissions. And I could not explore the solutions needed in island countries or in some of the world's poorest countries. But my hope is that the stories of scaling solutions in different political and financial circumstances, alongside building an ecosystem to support them, provides enough of a framework to think about how to shape the next few decades and get to zero emissions.

The forces needed to tackle climate change are finally starting to take shape. How far we can push them is up to those in power. If they don't act then it's unlikely the vast majority of people will sit still and suffer the consequences. A twelve-hour period in April 2019 in London shows why.

Sarah Breeden and Farhana Yamin had spent weeks preparing to voice their opinions on an issue they both cared about deeply. Each with over twenty years of experience in their respective professions, they were well prepared to deliver a call to action. But they were about to make their cases in very different ways.

Breeden's chosen venue was a conference room in a building in the City of London, the financial capital of the world. 'My message today is simple,' she began at a meeting organized by the Official Monetary and Financial Institutions Forum, a London-based think tank.[1] 'Climate change poses significant risks to the economy and to the financial system, and while these risks may seem abstract and far away, they are in fact very real, fast-approaching and in need of action today.'

Breeden was then, and still is, an executive director at the Bank of England, which manages the UK's currency and interest rates, regulates its banking system, and is the model for most national central banks in the modern world. Her remit at the time was overseeing the supervision of international banks that have a presence in London, which meant she worked with almost every important global financial institution. Her audience at the 15 April speech was made up of influential figures that control trillions of dollars worth of wealth: government ministers, investment bankers, asset managers and officials of financial ratings agencies.

Breeden's statements regarding climate change were bold for a central banker. 'Climate change is an unprecedented challenge,' she warned. 'I am sorry to say, there are no existing charts for us to follow.' The deaths of millions of people and the potential loss of trillions of dollars due to environmental damages were on the line. The climate crisis could halt the phenomenal progress humanity has made in the last 200 years and in the worst-case scenario, could even start to reverse it.

A central bank is supposed to ensure, as Breeden puts it, the 'safety and soundness' of financial institutions and shouldn't be concerned with saving the world. However, climate change poses real-world risks for financiers, such as when homes are made uninhabitable by rising sea levels and owners may be forced to default on their mortgage. It also poses business risks when it affects policy, such as when a government bans the use of coal and causes a publicly traded company to go bankrupt. These warnings have been made in one form or another for the past thirty years, but the fact that emissions have continued to rise show that few paid attention.

The twentieth-century American economist Hyman Minsky once noted that bankers, traders and other financiers periodically played the role of arsonist, setting the entire economy ablaze. The inferno isn't intended, of course, but is typically the result of a sudden collapse of one type of financial asset, for instance the stocks of fossil fuel companies. That then triggers the decline of other assets, such as the companies that make internal combustion engine cars, that turn oil into plastic or that own coal power plants. That, in turn, causes panic in the market and leads to sell-offs across the economy, triggering a recession.

It was clear to Breeden that some parts of the world were already on fire, both metaphorically and literally, and she wanted to use the financial system to put out the fire rather than throw fuel on it. She also knew that the financial system works best when the changes made to it are executed incrementally and slowly. As she said, it is like a group of 'supertankers rather than high-speed catamarans. To change course, therefore, we need early action, a sustained effort and a recognition that it is better to be roughly right now, not precisely right when it is too late.'

The very next day, just a few miles away, Farhana Yamin decided it was too late to make a reasoned case with the arsonists. She removed the top from a bottle of superglue and squeezed the contents onto her hands. She was standing outside the London headquarters of the oil giant Royal Dutch Shell.

Hours earlier, protestors had spray-painted *Shell Knew* and *Climate Criminals* on the Portland-stone-clad walls of the twenty-seven-storey Shell Centre. The police had arrested three protestors, but two other activists had managed to climb onto a glass ledge above the entrance and pull the ladder up with them. The police had cordoned off the area to prevent any more vandalism. After hours of negotiations, the two finally agreed to surrender and began climbing down from the ledge.

Yamin, who wasn't involved in the building-climbing stunt, used the distraction to make a run for it, and ducked under the police tape. A policeman caught her by the right arm; she went limp. As she fell the open palm of her left hand hit the floor and stuck there. 'I'm glued to the floor!' she shouted. The policeman let her go, and she then pressed her right palm to the floor, too.

A few feet away, her husband Michael and their son Rafi watched Yamin break the law. 'She's a lawyer,' Michael later told the press.[2] 'She believes in the rule of law. But non-violent, civil disobedience at some point becomes necessary.'

Yamin is not just any lawyer. She has spent twenty-five years helping craft international environmental laws such as the 1997 Kyoto Protocol, which established the first global carbon market, and the 2005 European Union Emissions Trading Scheme, one of the world's most progressive pieces of climate legislation. She has been a lead author of three reports by the Intergovernmental Panel on Climate Change, the highest global body on the subject. Her crowning glory she says is the inclusion of the net-zero emissions framework in the 2015 Paris Agreement, which was signed by every country in the world.

Now Yamin was glued to the floor outside Shell's offices, from where she made a statement to the world's press. 'Fossil fuel companies are ruining the legislative process internationally and have done everything in their power to stop governments from acting on climate change,' she said. 'Everybody who works here needs to wake up. Everybody who goes in and out of this building is treading on the fate of the world.'

It was a calculated move. Yamin knew she would probably be arrested and charged. 'That's why I'm here,' she said. 'In any court of law that I'm willing to submit to, any jury will find the truth: that these people are liars and they paid others to lie for them.' She also knew her act would make front-page news. She was, indeed, arrested. And then the next morning, after being released on bail – pending an investigation – BBC, Channel 4 and ITV all invited her to talk on their channels.

Yamin is one of the leaders of the activist group called Extinction Rebellion. In the week she was arrested outside Shell's London headquarters, authorities arrested more than 1,100 other protestors for disrupting the functioning of the British capital. Each protest site reflected one of Extinction Rebellion's four demands.

It began when activists pulled a pink boat with *Tell the Truth* painted on its hull across the busy pedestrian crossing of Oxford Circus. The boat, named *Berta Cáceres* after the Honduran environmental activist who was murdered in 2016, was used as a pulpit from which DJs played tunes and celebrities made speeches. It represented Extinction Rebellion's first demand: for governments to declare a climate emergency and tell the truth about the unfolding catastrophe.

At the same time near Marble Arch, at one end of Oxford Street, protestors blocked traffic with an inflated elephant that had *ecocide* written on it, reflecting Extinction Rebellion's second demand: to put an end to practices that are causing biodiversity loss.

Activists then drove a truck to Waterloo Bridge, one of the main routes in central London for crossing the River Thames, and used the vehicle to block the road while they transformed the bridge into a sort of garden. This was meant to signal Extinction Rebellion's third demand: for the UK to become carbon neutral by 2025.

In Parliament Square, outside the Palace of Westminster the rebels hosted lectures on climate science and screened David Attenborough's 2019 documentary, *Climate Change: The Facts*. This was an indication of the group's fourth demand: for governments to create citizens' assemblies to discuss how best to fight climate change.

When Yamin glued herself to the Shell building and, as a lawyer, chose to break the law, she embodied the frustration of the tens of thousands of people around the world – that an entrenched system prefers to take the path of least resistance, even in the face of overwhelming evidence. Yamin and the Extinction Rebellion activists had decided that they couldn't tolerate it anymore.

★

Back in the conference room, Breeden was getting into her stride, trying to generate institutional change from within. The risks from climate change, she said, 'are far-reaching in breadth and scope. They will affect all agents in the economy, in all sectors and across all geographies. The risks are eminently foreseeable. I cannot tell you now exactly what will happen and when.' But, she added, 'uncertainty about what will happen cannot lead to inaction and inertia'.

Breeden's words were made more powerful by the urgency in her speech. 'The size of those future risks will be determined by the actions we take today,' she continued. 'Climate change represents the tragedy of the horizon: by the time it is clear that climate change is creating risks that we want to reduce, it may already be too late to act.'

Much of humanity's economic progress over the past two centuries has been tied to fossil fuels; the link is still so direct that a rise in fossil fuel use remains one of the strongest factors in predicting a developing country's economic growth. But that pattern is starting to break down, with developed countries growing despite reducing fossil fuel use. Between 1990 and 2017 the UK economy grew by 60%, even though its carbon emissions declined by 40%.

Since the Paris Agreement the world's governments have begun to act. However, the scale of the changes needed is so big that private capital has to play its part. 'The need to act most obviously includes government through climate policy,' Breeden said. 'But since the financial risks that climate change creates are to be managed in all future states of the world, it is incumbent upon financial firms, and central banks and supervisors, to act too.'

Breeden was hopeful that there was still time to get on the right track. Having been at the Bank of England during the global financial crisis of 2008, and then seen the world recover, she had enough faith in her colleagues to believe that the system could do the right thing.

★

Yamin and the other protestors on the streets of London were also hopeful about the future of the world, but they put little stock in the system. Speeches made in private meetings to financiers would never be enough to trigger systemic change, they concluded; a louder call for action was needed.

'We are now facing an existential crisis. Humanity is standing at a crossroads,' declared Greta Thunberg in a speech at Marble Arch on Earth Day on 22 April 2019, as Extinction Rebellion protestors began to pack up protest gear spread across central London. The Swedish teenager is the founder of the School Strike for Climate movement, which began in 2018 and had grown to include more than 1.6 million schoolchildren around the world in less than six months when the kids marched across the world in March and April 2019. Their demand to those in power is simple: act like it's a crisis.

Breeden's speech encouraging wealthy financiers to start fighting climate change and Yamin's rebellion against an oil company that she believed has been blocking progress on climate action happened within hours of each other. The juxtaposition should not have been surprising, given that the two events both happened in the UK, a country that epitomizes the gap between what we are doing now and what we need to do to prevent a climate catastrophe. The UK was the first country in the world to set legally binding climate goals in 2008 and since then has done more than any other Western economy to cut carbon emissions. Every British political party is aligned in its support to fight climate change, unlike in countries including the US, Canada or Australia. And yet the UK government's Committee on Climate Change has been warning since 2019 that the country is not on track to hit its own climate goals, let alone the more ambitious ones set under the Paris climate agreement. If the UK is doing nowhere near enough, what hope is there for the rest of the globe?

Sarah Breeden wants to leverage big money to improve our lives and fight climate change. Farhana Yamin wants to change the status quo, because something is preventing the world from acting on the challenge.

Seen another way, Breeden and Yamin might look like two Davids going up against a Goliath but in fact, rather than defeating Goliath, they want to join forces with the giant to defeat an even bigger demon: climate change. The fight isn't for one group to stop the other, but for everyone to work together. This battle to create a new economic system – climate capitalism – will dominate the agenda for decades to come, and it is crucial that we understand it if we are to safeguard our planet for generations to come.

Given the disruption that's coming, the race to zero emissions is unlikely to be smooth. If history is a guide, it will probably be a series of missteps. Still, all the main forces needed to tackle this problem – politics, technology, finance – are headed in the right direction, with more and more people diverting their focus to the solutions. Time is running out and we need to pick up the pace.

Acknowledgements

It's my name on the cover, but this book would have been impossible without the contributions of so many. My deepest gratitude goes to all those who agreed to be interviewed – whether quoted or not – and for letting me investigate your ideas, companies or policies.

Elijah Wolfson, my editor at *Quartz*, was vital in helping me think through the book, the flow of words, the strength of the arguments and the depth of the reporting. I'm also thankful to Kevin Delaney, Jason Karaian and Gideon Lichfield at *Quartz*, for providing the space to experiment with telling the stories of climate solutions. I could not have written the China chapters without the help of Echo Huang and Beimeng Fu.

My colleagues at *Bloomberg News* proved tremendously helpful in giving me the support to finish the book through a hectic news schedule. The world experienced not just more intense climate impacts but also a pandemic and an economic shock to keep us all on our toes. Special thanks to Aaron Rutkoff, Sharon Chen, John Fraher and Will Kennedy for their encouragement. I also received reporting support from Eric Roston, Will Mathis and Kevin Crowley.

Thank you to Georgina Laycock, Joe Zigmond, Siam Hatzaw, Rosie Gailer, Alice Graham, Anna-Marie Fitzgerald and Caroline Westmore at John Murray Press, for taking a chance on the book. I'm also grateful to my agent Jonathan Conway for helping me stay calm over the four years it took to finish the writing, Mariyam

Haider for fact checking, Hilary Hammond for copy editing and Laurence Cole for proof reading. I greatly benefited from the inputs of Aaron Rutkoff, John Fraher, Will Kennedy, Jason Karaian, Frances Cairncross, Siddharth Singh, Monic Gupta, Daianna Karaian, Mun Keat Looi, Anne Kornahrens, Eric Roston, Will Mathis, Olivia Rudgard, Natasha White, Gautam Naik, Craig Trudell and Siobhan Wagner who read early drafts.

Most of all I would like to thank my wife Deeksha for being there at every moment of the process. Thanks also to my parents, Sangeeta and Hemant, for their unending support and encouragement and to my sister Surabhi for help with the cover.

Notes

Chapter 1: The Framework

1. See Madalina Ciobanu, 'Inside "The Race to Zero Emissions"', *Quartz*'s Latest In-depth Series and Newsletter about Carbon Capture', *Journalism*, 18 December 2017, https://www.journalism. co.uk/news/inside-the-race-to-zero-emissions-quartz-s-latest-in-depth-series-and-newsletter-about-carbon-capture/s2/a714857/

2. Alison Benjami, 'Stern: Climate Change a Market Failure', *Guardian*, 29 November 2007.

3. 'Ezra Klein Interviews Noam Chomsky', *The Ezra Klein Show*, podcast, 23 April 2021, https://www.nytimes.com/2021/04/23/ podcasts/ezra-klein-podcast-noam-chomsky-transcript.html

4. Marshall Burke, W. Matthew Davis and Noah S. Diffenbaugh, 'Large Potential Reduction in Economic Damages under UN Mitigation Targets', *Nature* 557 (2018): 549–53.

5. Even after accounting for the cost of climate action, a 2022 Deloitte report found that the net economic gain to the global economy would be more than $200 trillion over the next fifty years. 'Deloitte Research Reveals Inaction on Climate Change Could Cost the World's Economy US$178 Trillion by 2070', Deloitte, press release, 23 May 2022, https://www.deloitte.com/global/en/about/press-room/deloitte-research-reveals-inaction-on-climate-change-could -cost-the-world-economy-us-dollar-178-trillion-by-2070.html

6. Ann Gibbons, 'Experts Question Study Claiming to Pinpoint Birthplace of all Humans', *Science News*, 28 October 2019, doi: 10.1126/science.aba0155

7. Elizabeth Gamillo, 'Atmospheric Carbon Last Year Reached Levels Not Seen in 800,000 Years', *Science Insider*, 2 August 2018, doi: 10.1126/science.aau9866

8. Akshat Rathi, 'A 1912 News Article Ominously Forecasted the Catastrophic Effects of Fossil Fuels on Climate Change', *Quartz*, 24 October 2016, https://qz.com/817354/scientists-have-been-forecasting-that-burning-fossil-fuels-will-cause-climate-change-as-early-as-1882

9. John Vidal, 'Margaret Thatcher: An Unlikely Green Hero?', *Guardian*, 9 April 2013; Scott Waldman, 'Bush Had a Lasting Impact on Climate and Air Policy', *E&E News*, 3 December 2018, *Scientific American*, https://www.scientificamerican.com/article/bush-had-a-lasting-impact-on-climate-and-air-policy/

10. 'History of the IPCC', Intergovernmental Panel on Climate Change, https://www.ipcc.ch/about/history/

11. Kate Yoder, 'They Derailed Climate Action for a Decade. And Bragged about It', *Grist*, 15 April 2022, https://grist.org/accountability/how-the-global-climate-coalition-derailed-climate-action/

12. 'The Kyoto Protocol – Status of Ratification', UNFCCC (United Nations Climate Change), https://unfccc.int/process/the-kyoto-protocol/status-of-ratification

13. Helen Dewar and Kevin Sullivan, 'Senate Republicans Call Kyoto Pact Dead', *Washington Post*, 11 December 1997, https://www.washingtonpost.com/wp-srv/inatl/longterm/climate/stories/clim121197b.htm

14. Ana Swanson, 'How China used more cement in 3 years than the U.S. did in the entire 20th Century', *Washington Post*, 24 March 2015, https://www.washingtonpost.com/news/wonk/wp/2015/03/24/how-china-used-more-cement-in-3-years-than-the-u-s-did-in-the-entire-20th-century/

15. 'The Paris Agreement', UNFCCC (United Nations Climate Change), https://unfccc.int/process-and-meetings/the-paris-agreement

16. Will Mathis and Akshat Rathi, 'How Europe Ditched Russian Fossil Fuels with Spectacular Speed', *Bloomberg*, 21 February 2023,

https://www.bloomberg.com/news/features/2023-02-21/
ukraine-news-europe-ditches-russia-fossil-fuels-with-surprising
-speed

17. Saijel Kishan, 'There's $35 Trillion Invested in Sustainability, but $25 Trillion of That Isn't Doing Much', *Bloomberg*, 18 August 2021, https://www.bloomberg.com/news/articles/2021-08-18/-35-trillion-in-sustainability-funds-does-it-do-any-good

Chapter 2: The Bureaucrat

1. 'Number of Tesla Vehicles Delivered Worldwide from 1st Quarter 2016 to 4th Quarter 2022', Statista, https://www.statista.com/statistics/502208/tesla-quarterly-vehicle-deliveries/

2. Mark Kane, 'China: Plug-in Car Sales Increased by 75% in October 2022', *Inside EVs*, 27 November 2022, https://insideevs.com/news/623665/china-plugin-car-sales-october2022/

3. China also sells millions of low-speed electric vehicles each year, which can cost as little as $1,000 or around £800. But they are only available in China and thus are not included in the full EV figures. Akshat Rathi, 'The Cheapest Chinese Electric Cars are Coming to the US and Europe – for as Little as $9,000', *Quartz*, 4 February 2019, https://qz.com/1541380/the-cheapest-chinese-electric-cars-are-coming-to-the-us-and-europe/

4. Wan Gang in an interview as president of Tongji University in Shanghai. For a full transcript see https://www.shszx.gov.cn/node2/node1721/node1856/node1857/u1a16497.html

5. James P. Sterba, 'Peking Assessment Asserts Mao Made Errors as Leader', *New York Times*, 1 July 1981, https://www.nytimes.com/1981/07/01/world/peking-assessment-asserts-mao-made-errors-as-leader.html

6. 'The World's Leading Electric Car Visionary Isn't Elon Musk', *Bloomberg*, 26 September 2018, https://www.bloomberg.com/news/features/2018-09-26/world-s-electric-car-visionary-isn-t-musk-it-s-china-s-wan-gang

7. Lisa Margonelli, 'China's Next Cultural Revolution', *Wired*, 1 April 2005, https://www.wired.com/2005/04/china-4/

8. Levi Tillemann, *The Great Race: The Global Quest for the Car of the Future* (Simon & Schuster, 2015).

9. Dave Barthmuss, 'Who Ignored the Facts About the Electric Car?', GM Communications blog, 13 July 2006, http://www.altfuels.org/misc/onlygm.pdf

10. Jeremy Hodges, 'Electric Cars Are Cleaner Even When Powered by Coal', *Bloomberg*, 15 January 2019, https://www.bloomberg.com/news/articles/2019-01-15/electric-cars-seen-getting-cleaner-even-where-grids-rely-on-coal

11. May Zhou, 'China Drives up Global EV Sales to New Record', *China Daily Global*, 17 January 2023, https://www.chinadaily.com.cn/a/202301/17/WS63c602dba31057c47ebaa0ab.html

12. 'China's NEV Sales to Account for 20% of New Car Sales by 2025, 50% by 2035', Reuters, 27 October 2020, https://www.reuters.com/article/us-china-autos-electric-idUSKBN27C08C

13. Scott Kennedy, 'China's Risky Drive into New-Energy Vehicles', Centre for Strategic & International Studies, news page, 19 November 2018, https://www.csis.org/analysis/chinas-risky-drive-new-energy-vehicles

14. The total cost of owning an electric car, which includes its purchase price and lifetime cost of fuel, maintenance and taxes, is getting very close to that of a fossil fuel car. It is actually cheaper in many countries already. In the case of electric buses, the total cost of ownership is already cheaper than fossil fuel buses in almost all countries in the world.

15. Tom Taylor, 'IRA to Unlock Billions in EV Funding', EV Hub, 15 August 2022, https://www.atlasevhub.com/ira-to-unlock-billions-in-ev-funding/

16. Kate Abnett, 'EU Approves Effective Ban on New Fossil Fuel Cars from 2035', Reuters, 28 October 2022, https://www.reuters.com/markets/europe/eu-approves-effective-ban-new-fossil-fuel-cars-2035-2022-10-27/

17. Colin McKerracher, 'Phasing Out Europe's Combustion Engine Cars', Hyperdrive daily briefing, *Bloomberg*, 20 July 2021, https://

www.bloomberg.com/news/newsletters/2021-07-20/hyperdrive-daily-phasing-out-europe-s-combustion-engine-cars

18. 'World Oil Outlook', Organization of the Petroleum Exporting Countries, https://www.opec.org/opec_web/en/publications/340.htm

19. BloombergNEF, 'Electric Vehicle Outlook 2022', https://about.bnef.com/electric-vehicle-outlook/

20. 'VW Boosts Investment in Electric and Autonomous Car Technology to $86 Billion', Reuters, 13 November 2020, https://www.reuters.com/article/volkswagen-strategy-idUSKBN27T24O

21. Michael Wayland, 'GM Ups Spending on EVs and Autonomous Cars by 35% to $27 Billion', CNBC, 19 November 2020, https://www.cnbc.com/2020/11/19/gm-accelerating-ev-plans-with-additional-7-billion-announces-new-pickup.html

22. Sam Abuelsamid, 'Ford Doubling Investment in Electric Cars and Trucks to $22 Billion', *Forbes*, 4 February 2021, https://www.forbes.com/sites/samabuelsamid/2021/02/04/ford-doubles-investment-in-electrification-to-22b-7b-for-avs/?sh=2db2c23a2d25

23. 'Hyundai Commits $17 Billion to Add Electric, Driverless Cars', *Industry Week*, 4 December 2019, https://www.industryweek.com/technology-and-iiot/article/22028673/hyundai-commits-17-billion-to-add-electric-driverless-cars

Chapter 3: The Winner

1. Irene Preisinger and Victoria Bryan, 'China's CATL to Build Its First European EV Battery Factory in Germany', Reuters, 9 July 2018, https://www.reuters.com/article/us-bmw-catl-batteries-idUSKBN1JZ11Y. At the launch event in Berlin, Merkel was reported to have said: 'If we could do it ourselves, then I would not be upset.' An Alamy stock photo that captures the moment of the signing is available at https://bit.ly/4207r7s.

2. 'Information on Daimler AG', Daimler, https://www.daimler.com/company/tradition/company-history/1885-1886.html

3. https://edition.cnn.com/2021/01/28/business/toyota-volkswagen-japan-germany-intl-hnk/index.html

4. Michelle Toh, 'Toyota Overtakes Volkswagen as World's Biggest Automaker', CNN, 28 January 2021, https://qz.com/1582811/the-complete-guide-to-the-battery-revolution/

5. Christopher Jasper, 'New Electric Airplane to Make First Flight This Year', *Bloomberg*, 1 July 2021, https://www.bloomberg.com/news/articles/2021-07-01/eviation-s-electric-alice-plane-to-make-first-flight-this-year; 'World's Largest Electric Ferry Launches in Norway', Electrive, news page, 2 March 2021, https://www.electrive.com/2021/03/02/worlds-largest-electric-ferry-yet-goes-into-service-in-norway/

6. Alexander Yung, 'Germany Lags Behind Asia in E-Car Battery Race', *Der Spiegel*, 22 February 2019, https://www.spiegel.de/international/business/running-on-empty-germany-lags-behind-asia-in-e-car-battery-race-a-1254183.html

7. BloombergNEF battery cell manufacturers dataset, https://www.bnef.com/interactive-datasets/2d5d59acd9000002

8. For Alessandro Volta's first electrical battery see the Royal Institution, https://www.rigb.org/our-history/iconic-objects/iconic-objects-list/voltaic-pile; for more on voltaic electricity see http://ppp.unipv.it/collana/pages/libri/saggi/nuova%20voltiana3_pdf/cap4/4.pdf

9. Thanks to Shashank Sripad at Carnegie Mellon University for help with this calculation.

10. David Rand, 'History of Lead', Batteries International, news page, 21 September 2016, https://www.batteriesinternational.com/2016/09/21/history-of-lead/

11. Seth Fletcher, *Bottled Lightning* (Hill & Wang, 2011). Also Steve LeVine, *The Powerhouse* (Penguin, 2016).

12. Akshat Rathi, 'Winners of the 2019 Nobel Prize in Chemistry Developed Lithium-ion Batteries', *Quartz*, 9 October 2019, https://qz.com/1724449/nobel-prize-in-chemistry-winners-developed-lithium-ion-batteries/

13. Michael McCaul et al., 'Egregious Cases of Chinese Theft of American Intellectual Property', House Foreign Affairs Committee 2020, https://foreignaffairs.house.gov/wp-content/uploads/2020/02/Egregious-Cases-of-Chinese-Theft-of-American-Intellectual-Property.pdf

14. 'Olympics to Use 50 Li-ion Battery Powered Buses', Xinhua News Agency, 30 July 2007, http://www.china.org.cn/olympics/news/2007-07/30/content_1218985.htm

15. Jie Ma et al., 'The Breakneck Rise of China's Colossus of Electric-Car Batteries', *Bloomberg*, 1 February 2018, https://www.bloomberg.com/news/features/2018-02-01/the-breakneck-rise-of-china-s-colossus-of-electric-car-batteries

16. BMW stopped making the i3 in 2022 after a nine-year run: Greg Kable, 'BMW i3 to Cease Production in July after Nine Years', *Autocar*, 27 January 2022, https://www.autocar.co.uk/car-news/new-cars/bmw-i3-cease-production-july-after-nine-years

17. 'Elon Musk's China Battery Partner Is Now Richer Than Jack Ma', *Bloomberg*, 8 July 2021, https://www.bloomberg.com/news/articles/2021-07-08/elon-musk-s-battery-partner-in-china-is-now-richer-than-jack-ma

18. Bloomberg Billionaires Index, 4 June 2023, https://www.bloomberg.com/billionaires/

19. Akshat Rathi, 'Quantum Leap', *Bloomberg*, 14 April 2021, https://www.bloomberg.com/features/2021-quantumscape-battery/

20. Volkswagen was hit by Dieselgate in 2015, when it was confirmed that the company had cheated on its pollution control tests, and this scandal probably made it even more interested in electrification.

21. Jean Kumagai, 'Lithium Battery Recycling Finally Takes Off in North America and Europe', IEEE Spectrum, 5 January 2021, https://spectrum.ieee.org/lithiumion-battery-recycling-finally-takes-off-in-north-america-and-europe

22. 'Greenpeace Report Troubleshoots China's Electric Vehicles Boom, Highlights Critical Supply Risks for Lithium-ion Batteries', press release, Greenpeace, 30 October 2020, https://www.greenpeace.org/eastasia/press/6175/greenpeace-report-troubleshoots

-chinas-electric-vehicles-boom-highlights-critical-supply-risks-for-lithium-ion-batteries/

23. 'China's EV Battery Recycles to Peak in 2025, Requiring Policy Support', *Global Times*, 22 June 2021, https://www.globaltimes.cn/page/202106/1226776.shtml

24. Alejandra Salgado, 'China to Extend Battery-Metal Lead as Electric Cars Fuel Demand', *Bloomberg*, 30 June 2021, https://www.bloomberg.com/news/articles/2021-06-30/china-to-extend-battery-metal-lead-as-electric-cars-fuel-demand

25. Godehard Weyerer, 'New EU Laws Push Battery Recycling', *DW Business*, 18 June 2021, https://www.dw.com/en/battery-recycling-gains-speed-as-new-eu-regulation-pushes-investment/a-57933200

26. 'Council and Parliament Strike Provisional Deal to Create a Sustainable Life Cycle for Batteries', Council of the EU, press release, 9 December 2022, https://www.consilium.europa.eu/en/press/press-releases/2022/12/09/council-and-parliament-strike-provisional-deal-to-create-a-sustainable-life-cycle-for-batteries/

27. Zachary Shahan, 'The Really Big Battery Deal In The IRA That People Are Missing', *CleanTechnica*, 23 September 2022, https://cleantechnica.com/2022/09/23/the-really-big-battery-deal-in-the-ira-that-people-are-missing/

28. 'CATL Ups Investment in German Battery Plant', *ET Auto*, 30 June 2019, https://auto.economictimes.indiatimes.com/news/auto-components/catl-ups-investment-in-german-battery-plant/70008872

29. 'Battery Pack Prices Cited Below $100/kWh for the First Time in 2020, while Market Average Sits at $137/kWh', BloombergNEF, 16 December 2020, https://about.bnef.com/blog/battery-pack-prices-cited-below-100-kwh-for-the-first-time-in-2020-while-market-average-sits-at-137-kwh/

30. Colin McKerracher and Siobhan Wagner, 'At Least Two-Thirds of Global Car Sales Will Be Electric by 2040', *Bloomberg*, 9 August 2021, https://www.bloomberg.com/news/articles/2021-08-09/at-least-two-thirds-of-global-car-sales-will-be-electric-by-2040

31. Alexia Fernández Campbell, 'It Took 11 Months to Restore Power to Puerto Rico after Hurricane Maria. A Similar Crisis could Happen Again', *Vox*, 15 August 2018, https://www.vox.com/identities/2018/8/15/17692414/puerto-rico-power-electricity-restored-hurricane-maria

32. 'Puerto Rico Eyes Building the Energy Grid of the Future', *Quartz*, 13 September 2018, https://qz.com/1388117/puerto-rico-eyes-building-the-energy-grid-of-the-future/

33. 'SEforALL Analysis of SDG7 Progress – 2022', Sustainable Energy for All, 22 August 2022, https://www.seforall.org/data-stories/seforall-analysis-of-sdg7-progress

34. Danny Lee, 'Billionaire Vice Chairman of China Battery Giant CATL Resigns', *Bloomberg News*, 1 August 2022, https://www.bloomberg.com/news/articles/2022-08-01/billionaire-vice-chairman-of-china-battery-giant-catl-resigns

35. '2035 Electric Decarbonization Modeling Study', Berkeley Public Policy: The Goldman School, https://gspp.berkeley.edu/faculty-and-impact/centers/cepp/projects/2035-electric-decarbonization-modeling-study

36. David Roberts, 'Getting to 100% Renewables Requires Cheap Energy Storage. But How Cheap?', *Vox*, 20 September 2019, https://www.vox.com/energy-and-environment/2019/8/9/20767886/renewable-energy-storage-cost-electricity

37. Akshat Rathi, 'Why Rust Is the Future of Very Cheap Batteries', *Bloomberg*, 30 March 2023, https://www.bloomberg.com/news/features/2023-03-30/this-cheap-battery-can-power-green-energy-transition

38. Jess Shankleman and Akshat Rathi, 'China Vows Carbon Neutrality by 2060 in Major Climate Pledge', *Bloomberg*, 22 September 2020, https://www.bloomberg.com/news/articles/2020-09-22/china-pledges-carbon-neutrality-by-2060-and-tighter-climate-goal

Chapter 4: The Doer

1. S Bhuvaneshwari, 'Distress grips Pavagada, but there is no poll talk on water woes', *The Hindu*, 13 April 2019, https://www.thehindu.com/elections/lok-sabha-2019/distress-grips-pavagada-but-there-is-no-poll-talk-on-water-woes/article26824189.ece. Also, Arathi Menon, 'Given Land for Power, Pavagada Residents Now Powerless', *Mongabay*, 14 February 2022, https://india.mongabay.com/2022/02/given-land-for-power-pavagada-residents-now-powerless/

2. Michael Safi, 'Suicides of Nearly 60,000 Indian Farmers Linked to Climate Change, Study Claims', *Guardian*, 31 July 2017, https://www.theguardian.com/environment/2017/jul/31/suicides-of-nearly-60000-indian-farmers-linked-to-climate-change-study-claims

3. O. Hoegh-Guldberg et al., '2018: Impacts of 1.5 °C Global Warming on Natural and Human Systems', in *Global Warming of 1.5 °C: An IPCC Special Report*, ed. V. Masson-Delmotte et al. (Cambridge University Press, 2018), pp. 175–312, doi:10.1017/9781009157940.005

4. Shreya Jai, 'Solar Power at 4,000 Mw, Rajasthan in the Lead', *Business Standard*, 6 June 2015, https://www.business-standard.com/article/economy-policy/solar-power-capacity-touches-4-000-mw-rajasthan-races-ahead-of-gujarat-115060500726_1.html

5. Since the pandemic led to an abnormal decline in emissions across the world in 2020, it's best to compare countries based on 2019 figures before virus-related economic fluctuations began. Figure cited is based on data from the Global Carbon Project covering the period 1750 to 2018: Akshat Rathi and Archana Chaudhary, 'Modi Surprises Climate Summit with 2070 Net-Zero Vow for India', *Bloomberg*, 1 November 2021, https://www.bloomberg.com/news/articles/2021-11-01/india-will-reach-net-zero-emissions-by-2070-modi-tells-cop26

6. Akshat Rathi, 'India Will Have to Leapfrog Every Major Economy to Reach Net Zero by 2050', *Bloomberg*, 22 March 2021, https://www.bloomberg.com/news/articles/2021-03-22/india-will-have-to-leapfrog-every-major-economy-to-reach-net-zero-by-2050

7. Akshat Rathi, 'How to Think about India in a Net-Zero World', *Bloomberg*, 10 November 2020, https://www.bloomberg.com/news/articles/2020-11-10/india-does-not-hinder-climate-progress-without-net-zero-emissions-goal

8. '2020 the Fifth Costliest Year for Natural Disaster Losses (USD 201 bn)', UniCredit report, 21 January 2021, https://www.research.unicredit.eu/DocsKey/credit_docs_9999_179008.ashx?EXT=pdf&KEY=no3ZZLYZf5miJJA2_uTR8lNuOiLSXYTG63-PQtsM25g=&T=1

9. Akshat Rathi, 'All the Disasters in 2017 Were Even More Costly Than We Thought', *Quartz*, 11 April 2018, https://qz.com/1249867/global-disasters-in-2017-caused-337-billion-worth-of-economic-losses/

10. 'Western North American Extreme Heat Virtually Impossible Without Human-Caused Climate Change', World Weather Attribution, 7 July 2021, https://www.worldweatherattribution.org/western-north-american-extreme-heat-virtually-impossible-without-human-caused-climate-change/

11. 'Climate Change Likely Increased Extreme Monsoon Rainfall, Flooding Highly Vulnerable Communities in Pakistan', World Weather Attribution, 14 September 2022, https://www.worldweatherattribution.org/climate-change-likely-increased-extreme-monsoon-rainfall-flooding-highly-vulnerable-communities-in-pakistan/

12. Andrew Lee, 'Is Suzlon Back? India's Fallen Wind Power Star Unveils First Big Order in Years', *Recharge*, 1 June 2021, https://www.rechargenews.com/wind/is-suzlon-back-indias-fallen-wind-power-star-unveils-first-big-order-in-years/2-1-1018941

13. World Bank and International Finance Corporation, *Doing Business 2011: Making a Difference for Entrepreneurs* (International Bank for Reconstruction and Development/the World Bank, 2010).

14. 'Stocks That Fell Most since 2008 and Offer Good Investment', *Business Today*, 31 December 2011, https://www.businesstoday.in/magazine/stocks/story/stock-market-crash-worst-stocks-losers-25044-2011-12-01

15. Sergio Goncalves, 'EDP to Buy $2.2 bln U.S. Horizon Wind Energy', Reuters, 27 March 2007, https://www.reuters.com/article/us-edp-horizon-idUSL2715639720070327

16. 'Turning Around the Power Distribution Sector', August 2021, https://www.niti.gov.in/sites/default/files/2021-08/Electricity-Distribution-Report_030821.pdf

17. Ankur Mishra, 'India Heading for Another NPA Crisis? RBI Predicts Bad Loans Could Be as High as 12.5% by March', *Financial Express*, 25 July 2020, https://www.financialexpress.com/industry/banking-finance/india-heading-for-another-npa-crisis-rbi-predicts-bad-loans-could-be-as-high-as-12-5-by-march/2034493

18. 'Why India Can't Match the Gulf Region's Record-Low Solar Tariffs', Institute for Energy Economics and Financial Analysis, press release, 28 August 2020, https://ieefa.org/why-india-cant-match-the-gulf-regions-record-low-solar-tariffs/; Nithin Prasad, 'SECI's 2 GW Solar Auction Gets India a New Record-Low Tariff of ₹2.36/kWh', Mercom, 30 June 2020 or https://mercomindia.com/seci-solar-auction-india-record-low/

19. 'Vast Power of the Sun Is Tapped by Battery Using Sand Ingredient', *New York Times*, 26 April 1954, p. 1.

20. Vanessa Zainzinger, 'Breaking Efficiency Records with Tandem Solar Cells', Chemistry World, 27 April 2022, https://www.chemistryworld.com/news/breaking-efficiency-records-with-tandem-solar-cells/4015529.article

21. 'NREL Six-Junction Solar Cell Sets Two World Records for Efficiency', NREL, press release, 13 April 2020, https://www.nrel.gov/news/press/2020/nrel-six-junction-solar-cell-sets-two-world-records-for-efficiency.html

22. Fred Ferretti, 'The Way We Were: A Look Back at the Late Great Gas Shortage', *New York Times*, 14 April 1974.

23. Andrea Hsu, 'How Big Oil of the Past Helped Launch the Solar Industry of Today', *NPR*, 30 September 2019, https://www.npr.org/2019/09/30/763844598/how-big-oil-of-the-past-helped-launch-the-solar-industry-of-today

24. Geoffrey Jones and Loubna Bouamane, 'Power from Sunshine: A Business History of Solar Energy', Harvard Business School, working

paper, 25 May 2012, https://www.hbs.edu/ris/Publication%20Files/
12-105.pdf

25. Osamu Kimura and Tatsujiro Suzuki, '30 Years of Solar Energy
Development in Japan: Co-evolution Process of Technology,
Policies, and the Market', Central Research Institute of Electric
Power Industry, paper prepared for 2006 Berlin Conference on
Resource Policies: Effectiveness, Efficiency, and Equity, https://
citeseerx.ist.psu.edu/viewdoc/download?doi=10.1.1.454.8221&re
p=rep1&type=pdf

26. Aleh Cherp et al., 'Comparing Electricity Transitions: A Historical
Analysis of Nuclear, Wind and Solar Power in Germany and
Japan', *Energy Policy* 101 (February 2017): 612–28.

27. Kerstine Appunn, 'What's New In Germany's Renewable Energy
Act 2021', *Clean Energy Wire*, 23 April 2021, https://www.clean-
energywire.org/factsheets/whats-new-germanys-renewable-
energy-act-2021

28. 'Slashed Subsidies Send Shivers Through European Solar Industry',
Greenwire, 31 March 2010, https://archive.nytimes.com/www.
nytimes.com/gwire/2010/03/31/31greenwire-slashed-subsidies-
send-shivers-through-europea-32255.html?pagewanted=all

29. Huizhong Tan, 'Solar Energy in China: The Past, Present, and
Future', *China Focus*, 16 February 2021, https://chinafocus.ucsd.
edu/2021/02/16/solar-energy-in-china-the-past-present-and-
future/

30. 'Science: Sun Electricity', *Time*, 4 July 1955, https://content.time.
com/time/subscriber/article/0,33009,807289,00.html

31. 'Bloomberg New Energy Finance Says World Installed 132 GW
New Solar in 2020 & Forecasts 2021 Deployments to Shatter This
Number to Exceed 150 GW & Could End up at 194 GW', *Taiyang
News*, 19 January 2021, http://taiyangnews.info/business/bnef-132
-gw-solar-installed-globally-in-2020/

32. Government of India, National Solar Mission 2010, https://web.
archive.org/web/20180131105523/http://www.mnre.gov.in/solar-
mission/jnnsm/introduction-2/

33. 'Revision of Cumulative Targets Under National Solar Mission
from 20,000 MW by 2021–22 to 100,000 MW: India Surging

Ahead in the Field of Green Energy – 100 GW Solar Scale-Up Plan', Government of India, press release, 17 June 2015, https://pib.gov.in/newsite/PrintRelease.aspx?relid=122566

34. Simon Yuen, 'Indian Solar Capacity up 13.9 GW in 2022', *PV-Tech*, 17 March 2023, https://www.pv-tech.org/indian-solar-capacity-up-13-9gw-in-2022/

35. 'ReNew Power Signs India's first Round-the-Clock Renewable Energy PPA', ReNew Power, press release, August 2021, https://renewpower.in/wp-content/uploads/2021/08/ReNew_Power_PPA_SECI_RTC_V6-NJ.pdf

36. Nathaniel Bullard, 'It's Always Sunny in India's Renewable Power Market', *Bloomberg*, 4 June 2020, https://www.bloomberg.com/news/articles/2020-06-04/wind-plus-solar-power-means-a-renewable-boost-for-india-energy

37. Shreya Jai, 'ReNew Power to Invest $1.2 bn for Country's 1st Round-the-Clock RE Project', *Business Standard*, 8 August 2021, https://www.business-standard.com/article/companies/renew-power-to-invest-1-2-bn-for-country-s-1st-round-the-clock-re-project-121080700479_1.html

38. Mayank Aggarwal, 'Charted: The Biggest Hurdles for Narendra Modi's Solar Power Ambitions for India', *Quartz*, 27 May 2021, https://qz.com/india/2013255/the-hurdles-for-narendra-modis-solar-power-ambitions-for-india/

Chapter 5: The Fixer

1. Michael Ray, 'Paris Attacks of 2015', *Encyclopaedia Britannica*, 6 November 2022, https://www.britannica.com/event/Paris-attacks-of-2015

2. Since its inception, the IEA has also worked on 'energy conservation' or, as we would call it today, 'energy efficiency', namely lowering the amount of energy used without reducing the benefit gained. This was with a view towards energy security tied to fossil fuels, especially oil. Samuel VanVactor, 'Energy Conservation in the OECD: Progress and Results',

Journal of Energy and Development 3, no. 2 (spring 1978): 239–59.

3. The Paris-based organization's dues are paid separately from the OECD and are based on each member's share of oil imports as they stood in the 1970s. So even though it is now a net exporter of oil, the US continues to stump up the most cash, contributing about a quarter of the roughly €20 million annual budget.

4. The actual number varies from year to year, but in 2021 Ukraine's pipelines carried about 30% of Russia's total gas imports to Europe. Stuart Elliott, 'Russian Gas Flows into Europe Dip in April as Ukraine War Rumbles on', S&P Global, 6 May 2022, https://www.spglobal.com/commodityinsights/en/market-insights/latest-news/natural-gas/050622-russian-gas-flows-into-europe-dip-in-april-as-ukraine-war-rumbles-on

5. Fatih Birol, biography, European Parliament, meeting documents 2004–9, https://www.europarl.europa.eu/meetdocs/2004_2009/documents/dv/cv_birol_/cv_birol_en.pdf

6. See the membership page of the IEA's website, https://www.iea.org/about/membership

7. Marian Willuhn, 'Leaked: EU Hydrogen Strategy Eyes €140 Billion Turnover by 2030', *PV Magazine*, 19 June 2020, https://www.pv-magazine.com/2020/06/19/leaked-eu-hydrogen-strategy-eyes-e140-billion-turnover-by-2030/

8. William Wilkes and Vanessa Dezem, '"Climate Chancellor" Merkel Leaves Germans Flooded and Frustrated', *Bloomberg*, 23 July 2021, https://www.bloomberg.com/news/features/2021-07-23/angela-merkel-leaves-a-mixed-climate-legacy-in-germany

9. 'Biden–Harris Administration Launches American Innovation Effort to Create Jobs and Tackle the Climate Crisis', The White House, press release, 11 February 2021, https://www.whitehouse.gov/briefing-room/statements-releases/2021/02/11/biden-harris-administration-launches-american-innovation-effort-to-create-jobs-and-tackle-the-climate-crisis/

10. *The Future of Cooling*, IEA, May 2018, https://www.iea.org/reports/the-future-of-cooling

11. Rathi and Chaudhary, 'Modi Surprises Climate Summit'.

12. Jess Shankleman and Akshat Rathi, 'India's Last-Minute Coal Defense at COP26 Hid Role of China, U.S.', *Bloomberg*, 13 November 2021, https://www.bloomberg.com/news/articles/2021-11-13/india-s-last-minute-coal-defense-at-cop26-hid-role-of-china-u-s

13. Framework Convention on Climate Change, FCCC/CP/1996/2, Organizational Matters: Draft Rules of Procedure, 22 May 1996, https://unfccc.int/sites/default/files/resource/02_0.pdf

14. 'COP26: Alok Sharma Fights Back Tears as Glasgow Climate Pact Agreed', BBC News, 13 November 2021, https://www.bbc.co.uk/news/av/world-59276651

15. Luke Kemp, 'Votes not Vetoes', YouthPolicy, 16 April 2014, https://www.youthpolicy.org/blog/sustainability/unfccc-voting/

16. Akshat Rathi and Eric Roston, 'The World's Most Influential Energy Model Needs a Climate Update', *Bloomberg*, 29 May 2020, https://www.bloomberg.com/news/articles/2020-05-29/iea-s-world-energy-outlook-needs-a-1-5-c-climate-change-scenario

17. Lauri Myllyvirta, 'Why Does the IEA Keep Getting Renewables Wrong?', Unearthed, 14 November 2017, https://unearthed.greenpeace.org/2017/11/14/all-is-lost-renewable-energy-growth-will-hit-a-brick-wall-no-not-really-its-just-the-iea/

18. Simon Evans, 'New Fossil Fuels "Incompatible" with 1.5C Goal, Comprehensive Analysis Finds', Carbon Brief, 23 October 2022, https://www.carbonbrief.org/new-fossil-fuels-incompatible-with-1-5c-goal-comprehensive-analysis-finds/

19. David Wallace-Wells, 'How Big of a Climate Betrayal Is the Willow Oil Project?', *New York Times*, 16 March 2023, https://www.nytimes.com/2023/03/16/opinion/willow-oil-project-alaska-climate-change.html

Chapter 6: The Billionaire

1. 'Decarbonization Challenge for Steel', McKinsey & Company, 3 June 2020, https://www.mckinsey.com/industries/metals-and-mining/our-insights/decarbonization-challenge-for-steel

2. 'Climate Change and the Production of Iron and Steel', World Steel Association, policy paper, 2021, https://worldsteel.org/publications/policy-papers/climate-change-policy-paper/

3. Breakthrough Energy, about us page, https://www.breakthrough-energy.org/our-story/our-story

4. Eric Roston, Akshat Rathi and Christopher Cannon, 'How the World Is Spending $1.1 Trillion on Climate Technology', *Bloomberg*, 24 April 2023, https://www.bloomberg.com/graphics/2023-climate-tech-startups-where-to-invest/

5. Bill & Melinda Gates Foundation, fact sheet, https://www.gates-foundation.org/about/foundation-fact-sheet

6. Dylan Matthews, 'The Surprising Strategy Behind the Gates Foundation's Success', *Vox*, 11 February 2020, https://www.vox.com/future-perfect/2020/2/11/21133298/bill-gates-melinda-gates-money-foundation

7. Gavi, the Vaccine Alliance, about page, https://www.gavi.org/our-alliance/about

8. 'Bill and Melinda Gates Foundation: What Is It and What Does It Do?', BBC News, 4 May 2021, https://www.bbc.com/news/world-us-canada-56979480

9. Catherine Clifford, 'How Bill Gates' Company TerraPower Is Building Next-Generation Nuclear Power', CNBC, 8 April 2021, https://www.cnbc.com/2021/04/08/bill-gates-terrapower-is-building-next-generation-nuclear-power.html; David Blackmon, 'A Decade In Development, Liquid-Metal Batteries by Ambri May Soon Change the Energy Storage Game', *Forbes*, 2 September 2021, https://www.forbes.com/sites/davidblackmon/2021/09/02/bill-gates-backed-startup-might-change-the-renewable-energy-storage-game/?sh=5bf9e7e24a94; Katie Brigham, 'Bill Gates and Big Oil Back This Company That's Trying to Solve Climate Change by Sucking CO_2 Out of the Air', CNBC, 22 June 2019, https://www.cnbc.com/2019/06/21/carbon-engineering-co2-capture-backed-by-bill-gates-oil-companies.html

10. Rob Toews, 'Will This Generation of "Climate Tech" Be Different?', *Forbes*, 31 October 2021, https://www.forbes.com/

sites/robtoews/2021/10/31/will-this-generation-of-climate-tech-be-different/?sh=5b33997a4a62

11. Christina Binkley, 'Bill Gates Has a Master Plan for Battling Climate Change', *Wall Street Journal*, 15 February 2021, https://www.wsj.com/articles/bill-gates-interview-climate-change-book-11613173337

12. Leah McGrath Goodman, 'When a Billionaire Trader Loses His Edge', *Fortune*, 4 May 2012, https://fortune.com/2012/05/04/when-a-billionaire-trader-loses-his-edge/

13. Sam Apple, 'John Arnold Made a Fortune at Enron. Now He's Declared War on Bad Science', *Wired*, 22 January 2017, https://www.wired.com/2017/01/john-arnold-waging-war-on-bad-science/

14. Kevin Delaney, 'Bill Gates and Investors Worth $170 Billion Are Launching a Fund to Fight Climate Change Through Energy Innovation', *Quartz*, 11 December 2016, https://qz.com/859860/bill-gates-is-leading-a-new-1-billion-fund-focused-on-combatting-climate-change-through-innovation/

15. *Net Zero by 2050*, IEA, May 2021, https://www.iea.org/reports/net-zero-by-2050

16. 'Arpanet', Defense Advanced Research Projects Agency (DARPA), https://www.darpa.mil/about-us/timeline/arpanet

17. Akshat Rathi, 'Bill Gates-Led $1 Billion Fund Expands Its Portfolio of Startups Fighting Climate Change', *Quartz*, 26 August 2019, https://qz.com/1693546/breakthrough-energy-ventures-expands-its-portfolio-to-19-startups/

18. PwC, 'The State of Climate Tech 2022', https://www.pwc.com/gx/en/services/sustainability/publications/overcoming-inertia-in-climate-tech-investing.html, and 'The State of Climate Tech 2020', https://www.pwc.com/gx/en/services/sustainability/assets/pwc-the-state-of-climate-tech-2020.pdf

19. 'Concrete Needs to Lose Its Colossal Carbon Footprint', *Nature* 597 (2021): 593–4, doi: https://doi.org/10.1038/d41586-021-02612-5

20. European Cement Association (Cembureau), *The Role of Cement in the 2050 Low Carbon Economy*, 2013, https://cembureau.eu/

media/cpvoin5t/cembureau_2050roadmap_lowcarbonecon-
omy_2013-09-01.pdf

21. Leilac website, https://www.project-leilac.eu/leilac2-project;
    'LEILAC Project, Cutting-edge Technology to Efficiently Capture
    CO2', *Energy Industry Review*, 7 June 2021, https://energyindus-
    tryreview.com/construction/leilac-project-cutting-edge-technol-
    ogy-to-efficiently-capture-co2/

22. Akshat Rathi, 'Bill Gates-Led Fund Invests in European Green
    Cement Maker', *Bloomberg*, 10 May 2021, https://www.bloomb-
    erg.com/news/articles/2021-05-10/bill-gates-led-fund-invests-in
    -european-green-cement-startup

23. Barry O'Halloran, 'Ecocem Plans €45m "Green" Cement Mill in
    US Despite Opposition', *Irish Times*, 25 April 2019, https://www.
    irishtimes.com/business/manufacturing/ecocem-plans-45m-
    green-cement-mill-in-us-despite-opposition-1.3870599

24. Akshat Rathi, 'The Material That Built the Modern World Is Also
    Destroying It. Here's a Fix', *Quartz*, 6 December 2017, https://
    qz.com/1123875/the-material-that-built-the-modern-world-is-
    also-destroying-it-heres-a-fix/

25. 'Shrinking Carbon Emissions Through Innovative Cement and
    Concrete Technologies', webinar, Carbon Cure, https://www.
    carboncure.com/resources/shrinking-carbon-emissions-through-
    innovative-cement-and-concrete-technologies/

26. Large polluters in the European Union are bound by the Emissions
    Trading Scheme, which sets a cap on each polluter's annual pollu-
    tion. Those that produce fewer emissions than the cap can trade
    the allowance with those who produce higher emissions. A Cap-
    and-trade system, like a direct tax on carbon pollution, also incen-
    tivizes polluters to emit less carbon dioxide.

27. Office of Nuclear Energy, 'U.S. Department of Energy Announces
    $160 Million in First Awards under Advanced Reactor
    Demonstration Program', press release, 13 October 2020, https://
    www.energy.gov/ne/articles/us-department-energy-announces-
    160-million-first-awards-under-advanced-reactor

28. Will Mathis and Akshat Rathi, 'Better Cables Could Halve U.S.
    Grid Emissions by 2030, Gates-Led Group Says', *Bloomberg*, 2

March 2021, https://www.bloomberg.com/news/articles/2021-03-02/bill-gates-led-group-shows-u-s-grid-emissions-can-fall-45

29. Akshat Rathi, 'Grant From Bill Gates-led Fund Will Make Green Jet Fuel As Cheap As Fossil Fuels', *Bloomberg*, 19 October 2022, https://www.bloomberg.com/news/articles/2022-10-19/bill-gates-led-fund-makes-green-jet-fuel-as-cheap-as-oil-based-competitor

30. Henry Sanderson, 'QuantumScape: Can Battery Pioneer Live up to the Hype?', *Financial Times*, 20 January 2021, https://www.ft.com/content/c31ca3ce-5e83-452c-86cb-3d1646490c7a

Chapter 7: The Wrangler

1. 'Sleipner Partnership Releases CO2 Storage Data', Equinor, 12 June 2019, https://www.equinor.com/news/archive/2019-06-12-sleipner-co2-storage-data

2. Akshat Rathi, 'How to Think about Negative Emissions in the Climate Fight', *Bloomberg*, 13 April 2021, https://www.bloomberg.com/news/articles/2021-04-13/how-to-think-about-negative-emissions-in-the-climate-fight

3. R. Stuart Haszeldine, 'Carbon Capture and Storage: How Green Can Black Be?', *Science* 325, no. 5948 (September 2009): 1747–52, doi: 10.1126/science.1172246

4. 'CO2 Enhanced Oil Recovery', Global Energy Monitor, 19 July 2021, https://www.gem.wiki/CO2_enhanced_oil_recovery

5. 'The Norwegian State as Shareholder', Equinor, https://www.equinor.com/about-us/the-norwegian-state-as-shareholder

6. Akshat Rathi, 'The World's First "Negative Emissions" Plant Has Begun Operation – Turning Carbon Dioxide into Stone', *Quartz*, 12 October 2017, https://qz.com/1100221/the-worlds-first-negative-emissions-plant-has-opened-in-iceland-turning-carbon-dioxide-into-stone

7. 'Open Letter to Christiana Figueres, Executive Secretary of the United Nations Framework Convention on Climate Change',

Scottish Carbon Capture and Storage, 2015, https://www.sccs.org.uk/cop21-open-letter

8. Angely Mercado, 'In Memory of @BPDeezNutzz, the Best Greentrolling Account to Ever Grace Twitter', Gizmodo, 22 November 2022, https://gizmodo.com/bpdeeznutzz-twitter-account-suspended-1849802206

9. Hannah Ritchie and Max Roser, 'Electricity Mix', Our World In Data, 2022, https://ourworldindata.org/electricity-mix

10. US Office of Fossil Energy and Carbon Management, 'Fossil Energy Budget Request for Fiscal Year 2013', 27 March 2012, https://www.energy.gov/fecm/articles/fossil-energy-budget-request-fiscal-year-2013

11. US Department of Energy, 'President Trump Releases FY 2021 Budget Request', 10 February 2020, https://www.energy.gov/articles/president-trump-releases-fy-2021-budget-request

12. 'Kemper County IGCC Fact Sheet: Carbon Dioxide Capture and Storage Project', Carbon Capture and Sequestration Technologies Program, September 2016, https://sequestration.mit.edu/tools/projects/kemper.html

13. Ian Urbina, 'Piles of Dirty Secrets Behind a Model "Clean" Coal Project, *New York Times*, 5 July 2016, https://www.nytimes.com/2016/07/05/science/kemper-coal-mississippi.html?_r=0

14. Katie Fehrenbacher, 'Carbon Capture Suffers a Huge Setback as Kempler Plant Suspends Work', GTM, 29 June 2017, https://www.greentechmedia.com/articles/read/carbon-capture-suffers-a-huge-setback-as-kemper-plant-suspends-work

15. Akshat Rathi, 'Humanity's Fight against Climate Change Is Failing. One Technology Can Change That', *Quartz*, 4 December 2017, https://qz.com/1144298/humanitys-fight-against-climate-change-is-failing-one-technology-can-change-that

16. 'Petra Nova W.A. Parish Fact Sheet: Carbon Dioxide Capture and Storage Project', Carbon Capture and Sequestration Technologies Program, 30 September 2016, https://sequestration.mit.edu/tools/projects/wa_parish.html

17. Rathi, 'Humanity's Fight'.

18. Kevin Crowley, 'The World's Largest Carbon Capture Plant Gets a Second Chance in Texas', *Bloomberg*, 8 February 2023, https://www.bloomberg.com/news/articles/2023-02-08/the-world-s-largest-carbon-capture-plant-gets-a-second-chance-in-texas?sref=jjXJRDFv#xj4y7vzkg

19. Benjamin, 'Stern: Climate Change a "Market Failure"'.

20. 'William D. Nordhaus – Facts – 2018', NobelPrize, 10 May 2023, https://www.nobelprize.org/prizes/economic-sciences/2018/nordhaus/facts/

21. 'Emissions to Air', Norwegian Petroleum, 6 October 2022, https://www.norskpetroleum.no/en/environment-and-technology/emissions-to-air/

22. 'British Carbon Tax Leads to 93% Drop in Coal-fired Electricity', UCL News, 27 January 2020, https://www.ucl.ac.uk/news/2020/jan/british-carbon-tax-leads-93-drop-coal-fired-electricity

23. John M. Broder, ' "Cap and Trade" Loses Its Standing as Energy Policy of Choice', *New York Times*, 26 March 2010, https://www.nytimes.com/2010/03/26/science/earth/26climate.html

24. Akshat Rathi, 'A Republican Group is Framing Its Proposed Carbon Tax as "Environmental Insurance," Not a Tax', *Quartz*, 8 February 2017, https://qz.com/905688/a-republican-group-making-the-case-for-a-carbon-tax-to-donald-trumps-administration-needs-to-just-look-at-what-happened-in-australia

25. Kyle Bakx, 'The Big Election Winner? The Carbon Tax', CBC News, 22 October 2019, https://www.cbc.ca/news/business/trudeau-sheer-election-carbon-tax-1.5330829

26. Kim Willsher, 'Macron Scraps Fuel Tax Rise in Face of *gilets jaunes* Protests', *Guardian*, 5 December 2018, https://www.theguardian.com/world/2018/dec/05/france-wealth-tax-changes-gilets-jaunes-protests-president-macron

27. 'Section 45Q Credit for Carbon Oxide Sequestration', IEA, policy document, 4 November 2022, https://www.iea.org/policies/4986-section-45q-credit-for-carbon-oxide-sequestration

28. Kelemen recounted this story at Columbia University's 2017 annual energy summit: see Center on Global Energy Policy, '2017

Annual Energy Summit: Part 6', YouTube, 13 April 2017, https://www.youtube.com/watch?v=fgTcG3dXmFQ

29. Matthew E. Kahn et al., 'Long-term Macroeconomic Effects of Climate Change', International Monetary Fund, working paper, 11 October 2019, https://www.imf.org/en/Publications/WP/Issues/2019/10/11/Long-Term-Macroeconomic-Effects-of-Climate-Change-A-Cross-Country-Analysis-48691

30. 'Absolute Impact: Why Oil and Gas Companies Need Credible Plans to Meet Climate Targets', Carbon Tracker, 12 May 2022, https://carbontracker.org/reports/absolute-impact-2022/

Chapter 8: The Reformer

1. Stacey Wong, Oscar Boyd and Akshat Rathi, 'Transcript *Zero* Episode 7: Would You Buy "Net-Zero Oil"', *Bloomberg*, 27 October 2022, https://www.bloomberg.com/news/articles/2022-10-27/transcript-zero-episode-7-would-you-buy-net-zero-oil

2. Since the list was made there have been two mergers – Exxon and Mobil became ExxonMobil, and ChevronTexaco and Texaco became Chevron – and Amoco was acquired by BP. Even after accounting for these changes, there would have been at least six oil companies in the Fortune 500's top twenty in the 1990 list.

3. For a database of fifty years of Fortune's list of America's largest corporations see CNN Money, https://money.cnn.com/magazines/fortune/fortune500_archive/full/1990/. For a list of Fortune 500 companies and their websites see Zyxware Technologies, https://www.zyxware.com/articles/4344/list-of-fortune-500-companies-and-their-websites

4. Daniel Yergin, *The Prize* (Simon & Schuster, 1990).

5. Yergin, *The Prize*.

6. Vicki Hollub, The Bob and Elizabeth Dole Series on Leadership, Bipartisan Policy Center, 14 December 2018. https://www.youtube.com/watch?v=r5yo1Oi_Ixw

7. Akshat Rathi, 'Vicki Hollub Is Showing Big Oil How to Survive Climate Change', *Quartz*, 1 July 2019, https://qz.com/1641227/vicki-hollub-is-showing-big-oil-how-to-survive-climate-change

8. Jonathan Watts, Garry Blight, Lydia McMullan and Pablo Gutiérrez, 'Half a century of dither and denial – a climate crisis timeline', *The Guardian*, 9 October 2019, https://www.theguardian.com/environment/ng-interactive/2019/oct/09/half-century-dither-denial-climate-crisis-timeline

9. Global Climate Coalition, *DeSmog*, 2019. https://www.desmog.com/global-climate-coalition/

10. Kate Yoder, 'Big Oil spent $3.6 billion to clean up its image, and it's working', *Grist*, 24 December 2019, https://grist.org/energy/big-oil-spent-3-6-billion-on-climate-ads-and-its-working/

11. Kevin Crowley and Akshat Rathi, 'Exxon Holds Back on Technology That Could Slow Climate Change', *Bloomberg*, 7 December 2020, https://www.bloomberg.com/news/features/2020-12-07/exxon-s-xom-carbon-capture-project-stalled-by-covid-19

12. Pope Francis, Address to executives of energy industry, 9 June 2018, https://www.vatican.va/content/francesco/en/speeches/2018/june/documents/papa-francesco_20180609_imprenditori-energia.html

13. Akshat Rathi, 'A Tiny Tweak in California Law Is Creating a Strange Thing: Carbon-Negative Oil', *Quartz*, 1 July 2019, https://qz.com/1638096/the-story-behind-the-worlds-first-large-direct-air-capture-plant

14. Kevin Crowley and Ari Natter, 'Manchin Spurs US Reversal on Carbon Capture Funding in Win for Big Oil', *Bloomberg*, 14 December 2022, https://www.bloomberg.com/news/articles/2022-12-14/manchin-spurs-us-reversal-on-carbon-capture-funding-in-win-for-big-oil

15. Wong, Boyd and Rathi, 'Transcript *Zero* Episode 7'.

16. Ron Bousso, 'Shell Considers Exiting UK, German, Dutch Energy Retail Businesses', Reuters, 26 January 2023, https://www.reuters.com/business/energy/shell-considers-exiting-uk-german-dutch-energy-retail-businesses-2023-01-26/

17. Akshat Rathi, 'Musk's $100 Million Prize Is for Tech the World Desperately Needs', *Bloomberg*, 22 January 2021, https://www.bloomberg.com/news/articles/2021-01-22/musk-s-100-million-prize-is-for-tech-the-world-desperately-needs

18. Akshat Rathi, 'Stripe, Alphabet and Others to Spend Nearly $1 Billion on Carbon Removal', *Bloomberg*, 12 April 2022, https://www.bloomberg.com/news/articles/2022-04-12/stripe-alphabet-meta-join-to-fund-carbon-removal

Chapter 9: The Enforcer

1. 'Reinventing the Business Model: Leading in the New Landscape', Corporate Research Forum, 2021, https://www.crforum.co.uk/wp-content/uploads/2021/05/Reinventing-the-Business-Model-PMNs.pdf

2. David Laister, 'Why Theresa May Picked Danish Energy Giant Ørsted for Energy Speech in Grimsby', *Grimsby Telegraph*, 8 March 2019.

3. Mogens Rüdiger, 'The 1973 Oil Crisis and the Designing of a Danish Energy Policy', *Historical Social Research* 39, no. 4 (2014): 94–112.

4. 'From Black to Green', State of Green, press release, 31 May 2021, https://stateofgreen.com/en/news/from-black-to-green-a-state-owned-energy-companys-shift-and-the-framework-that-made-it-possible/

5. *Regulation and Planning of District Heating in Denmark*, Danish Energy Agency, June 2017, https://ens.dk/sites/ens.dk/files/Globalcooperation/regulation_and_planning_of_district_heating_in_denmark.pdf

6. Brian Motherway, 'Energy Efficiency Is the First Fuel, and Demand for It Needs to Grow', IEA, 19 December 2019, https://www.iea.org/commentaries/energy-efficiency-is-the-first-fuel-and-demand-for-it-needs-to-grow

7. Will Mathis, 'Inventer of Wind Turbine is Trying to Harness Unlimited Power', *Bloomberg*, 5 June 2020, https://www.

bloomberg.com/news/features/2020-06-05/floating-wind-farms-could-supply-the-world-s-electricity-by-2040?sref=jjXJRDFv

8. Henrik Stiesdal, 'From Herborg Blacksmith to Vestas', in *Wind Power for the World: The Rise of Modern Wind Energy*, ed. Preben Maegaard, Anna Krenz and Wolfgang Palz (Routledge, 2013).

9. Mette Fraende and Geert de Clercq, 'Goldman Sachs Funds Invest in DONG Energy, Seek IPO', Reuters, 3 October 2013, https://www.reuters.com/article/us-dong-investors-idUSBRE9920L120131003

10. Julian Spector, 'Dong Energy Divests Its Oil and Gas Businesses to Focus on Renewables', GTM, 26 May 2017, https://www.greentechmedia.com/articles/read/dong-energy-divests-upstream-oil-and-gas-business-to-focus-on-renewables

Chapter 10: The Campaigner

1. 'Climate Change Bill – Third Reading (and other amendments) – 28 October 2008', The Public Whip, https://www.publicwhip.org.uk/division.php?date=2008-10-28&number=298&display=all possible

2. The Climate Change Act (2008), Institute for Government, https://www.instituteforgovernment.org.uk/sites/default/files/climate_change_act.pdf

3. 'A Burnt-out Case', *The Economist*, 9 April 1998, https://www.economist.com/britain/1998/04/09/a-burnt-out-case

4. Nicholas Stern, 'The Economics of Climate Change: The Stern Review', 30 October 2006, https://www.lse.ac.uk/granthaminstitute/publication/the-economics-of-climate-change-the-stern-review/

5. 'The UK Climate Change Act', CCC Insights Briefing 1, Climate Change Committee, October 2020, https://www.theccc.org.uk/wp-content/uploads/2020/10/CCC-Insights-Briefing-1-The-UK-Climate-Change-Act.pdf

6. Climate Change Act of 2008, First Reading, 14 November 2007, https://publications.parliament.uk/pa/ld200708/ldhansrd/text/71114-0002.htm#07111435000003

7. The Global Climate Legislation Study, 2015, https://www.lse.ac.uk/GranthamInstitute/wp-content/uploads/2015/05/Global_climate_legislation_study_20151.pdf

8. *Urgenda Foundation* v. *State of the Netherlands* [2015] HAZA C/09/00456689.

9. Neubauer, et al. v. Germany, 2020, http://climatecasechart.com/non-us-case/neubauer-et-al-v-germany/

10. 'Germany Raises Ambition to Net Zero by 2045 after Landmark Court Hearing', *Climate Change News*, 5 May 2021, https://climatechangenews.com/2021/05/05/germany-raises-ambition-net-zero-2045-landmark-court-ruling/

11. *Sharma and others* v. *Minister for the Environment* [2021] FCA 560 and FCA 774; *Sharma* v. *Minister for the Environment* [2022] FCAFC 35, at [7].

12. For a full list see the online database Climate Change Laws of the World, https://climate-laws.org/

13. Amy Gunia, 'A Handful of Climate-Focused Independents Just Upended Australia's Political System. Here's What Comes Next', *Time*, 25 May 2022, https://time.com/6181345/climate-independents-australia-election/

14. *R (On the Application of) Friends of the Earth Ltd & Ors* v. *Secretary of State for Business, Energy and Industrial Strategy* [2022] EWHC 1841 (Admin). http://climatecasechart.com/non-us-case/r-oao-friends-of-the-earth-v-secretary-of-state-for-business-energy-and-industrial-strategy/

15. Steven Pearlstein, 'Will America's woes bring down democracy and capitalism worldwide?', *Washington Post*, 9 February 2023, https://www.washingtonpost.com/books/2023/02/09/capitalism-crisis-book-martin-wolf/

Chapter 11: The Capitalist

1. For the most recent GlobeScan Sustainability Survey see https://globescan.com/2022/06/23/2022-sustainability-leaders-report/

2. 'Unilever at a Glance', https://www.unilever.com/our-company/at-a-glance/

3. 'Unilever plc', Google Finance, https://www.google.com/finance/quote/ULVR:LON?window=MAX. According to Unilever's annual reports, in 2009 revenues stood at 39.8 billion euros (https://www.unilever.com/Images/ir-unilever-ar09_tcm244-421759_en.pdf) and in 2020 at 52 billion euros (https://www.unilever.com/Images/unilever-annual-report-and-accounts-2019_tcm244-547893_en.pdf). Figures available for Scope 1 and 2 emissions were for 2013 and 2019.

4. Matt Egan, 'Exxon Was the World's Largest Company in 2013. Now It's Being Kicked Out of the Dow', CNN Business, 25 August 2020, https://edition.cnn.com/2020/08/25/investing/exxon-stock-dow-oil/index.html

5. Kevin Crowley and Javier Blas, 'Exxon Defends Dividend after Posting First Annual Loss in Decades', *Bloomberg*, 2 February 2021, https://www.bloomberg.com/news/articles/2021-02-02/exxon-s-19-billion-writedown-caps-first-annual-loss-in-40-years

6. Jennifer Hiller and Svea Herbst-Bayliss, 'Little Engine No. 1 Beat Exxon with Just $12.5 mln – Sources', Reuters, 29 June 2021, https://www.reuters.com/business/little-engine-no-1-beat-exxon with just 125-mln sources 2021-06-29/

7. 'Carbon Emissions of Richest 1 Percent More Than Double the Emissions of the Poorest Half of Humanity', Oxfam, press release, 21 September 2020, https://www.oxfam.org/en/press-releases/carbon-emissions-richest-1-percent-more-double-emissions-poorest-half-humanity

8. Daniel J. Fiorino, *Can Democracy Handle Climate Change?* (Wiley, 2018).

9. Christian Davies, 'Natural Disasters Drive North Korea's Embrace of International Climate Goals', *Financial Times*, 22 January 2022, https://www.ft.com/content/d637c465-fc9e-4254-8191-193ac5eae30e

10. In *Can Democracy Handle Climate Change?* (Wiley, 2018), American professor Daniel Fiorino shows that robust democracy combined with the innovative power of the private sector is the best bet for tackling climate change.

NOTES

11. Bill Snyder, 'Unilever CEO: Refocus Your Ambitions', Stanford Business, Insights, 21 June 2016, https://www.gsb.stanford.edu/insights/unilever-ceo-refocus-your-ambitions

12. *New York Times*, 29 August 2019, https://www.nytimes.com/2019/08/29/business/paul-polman-unilever-corner-office.html

13. Lee Romney, 'Workers' Loyalty to Lever Bros. Outlasted Their Jobs', *LA Times*, 1 January 2000, https://www.latimes.com/archives/la-xpm-2000-jan-01-me-49704-story.html

14. Jasper Jolly, 'Unilever Workers Will Never Return to Desks Full-time, Says Boss', *Guardian*, 13 January 2021, https://www.theguardian.com/business/2021/jan/13/unilever-workers-will-never-return-to-desks-full-time-says-boss

15. Erik Meijaard et al., 'The Environmental Impacts of Palm Oil in Context', *Nature Plants* 6 (2020): 1418–26; IUCN, https://www.iucn.org/resources/issues-briefs/palm-oil-and-biodiversity#issue

16. 'Sustainable and Deforestation-Free Palm Oil', Unilever, https://www.unilever.com/planet-and-society/protect-and-regenerate-nature/sustainable-palm-oil/

17. Pablo Robles et al., 'The World's Addiction to Palm Oil Is only Getting Worse', *Bloomberg*, 8 November 2021, https://www.bloomberg.com/graphics/2021-palm-oil-deforestation-climate-change/

18. *Unilever Sustainable Living Plan 2010 to 2020*, March 2021, https://assets.unilever.com/files/92ui5egz/production/16cb778e4d31b-81509dc5937001559f1f5c863ab.pdf

19. Antoine Gara, 'Kraft Heinz Withdraws Its $143 Billion Bid for Unilever', *Forbes*, 19 February 2017, https://www.forbes.com/sites/antoinegara/2017/02/19/kraft-heinz-withdraws-its-143-billion-bid-for-unilever/?sh=773d35f54063

20. 'Statement Regarding Announcement by the Kraft Heinz Company of a Potential Transaction', Unilever, press release, 17 February 2017, https://www.unilever.com/news/press-and-media/press-releases/2017/statement-regarding-announcement-by-the-kraft-heinz-company/

21. Thomas Buckley and Matthew Campbell, 'If Unilever Can't Make Feel-Good Capitalism Work, Who Can?', *Bloomberg*, 31 August

2017, https://www.bloomberg.com/news/features/2017-08-31/if-unilever-can-t-make-feel-good-capitalism-work-who-can

22. Arash Massoudi and James Fontella-Khan, 'The $143bn Flop: How Warren Buffett and 3G Lost Unilever', *Financial Times*, 21 February 2017, https://www.ft.com/content/d846766e-f81b-11e6-bd4e-68d53499ed71

23. 'After Protests, Unilever Does About-Face on Palm Oil', *Wall Street Journal*, 2 May 2008, https://www.wsj.com/articles/SB1209667324426660143

24. Thomas Buckley and Matthew Campbell, 'If Unilever Can't Make Feel-Good Capitalism Work, Who Can?'.

25. 'Collectives', Imagine website, https://imagine.one/collectives/

26. Matt Levine, 'Exxon Lost a Climate Proxy Fight', *Bloomberg*, 27 May 2021, https://www.bloomberg.com/opinion/articles/2021-05-27/exxon-lost-a-climate-proxy-fight

27. ExxonMobil, https://www.ourenergypolicy.org/wp-content/uploads/2018/03/2018-Energy-and-Carbon-Summary.pdf; Kathy Hipple and Tom Sanzillo, 'ExxonMobil's Climate Risk Report: Defective and Unresponsive', Institute for Energy Economics and Financial Analysis (IEEFA), March 2018, https://ieefa.org/wp-content/uploads/2018/03/ExxonMobils-Climate-Risk-Report-Defective-and-Unresponsive-March-2018.pdf

28. Akshat Rathi and Alastair Marsh, 'Church of England Unloads Exxon Shares on Failed Emission Goals', *Bloomberg*, 8 October 2020, https://www.bloomberg.com/news/articles/2020-10-08/church-of-england-pensions-board-has-divested-from-exxonmobil

29. Akshat Rathi, 'Bill Gates Shows How Hard It Can Be to Divest from Fossil Fuels', *Bloomberg*, 15 February 2021, https://www.bloomberg.com/news/articles/2021-02-15/bill-gates-in-new-climate-book-talks-about-finally-divesting-from-oil

30. Luigi Zingales and Bethany McLean, 'The Engine No. 1, David vs Exxon Goliath, with Chris James', *Capitalisn't* podcast, 15 July 2021, https://www.capitalisnt.com/episodes/the-engine-no-1-david-vs-exxon-goliath-with-chris-james

31. Saijel Kishan and Joe Carroll, 'The Little Engline That Won an Environmental Victory over Exxon', *Bloomberg*, 9 June 2021, https:

//www.bloomberg.com/news/articles/2021-06-09/engine-no-1
-proxy-campaign-against-exxon-xom-marks-win-for-esg
-activists

32. Matt Phillips, 'Exxon's Board Defeat Signals the Rise of Social-
Good Activists', *New York Times*, 9 June 2021, https://www.
nytimes.com/2021/06/09/business/exxon-mobil-engine-no1-
activist.html

33. David McLaughlin and Annie Massa, 'The Hidden Dangers of the
Great Index Fund Takeover', *Bloomberg*, 9 January 2020, https://
www.bloomberg.com/news/features/2020-01-09/the-hidden-
dangers-of-the-great-index-fund-takeover

34. *S&P 500: The Gauge of the U.S, Large-Cap Market*, S&P Dow
Jones Indices, https://www.spglobal.com/spdji/en/documents/
additional-material/sp-500-brochure.pdf

35. Annie Lowrey, 'Could Index Funds Be "Worse Than Marxism"?',
The Atlantic, 5 April 2021, https://www.theatlantic.com/ideas/
archive/2021/04/the-autopilot-economy/618497/

36. Alastair Marsh, 'Climate Activist Who Took on BlackRock Now
Takes Aim at Vanguard', *Bloomberg*, 3 March 2021, https://www.
bloomberg.com/news/articles/2021-03-03/climate-activist-casey
-harrell-took-on-blackrock-takes-aim-at-vanguard

37. Jessica Camille Aguirre, 'The Little Hedge Fund Taking Down
Big Oil', *New York Times*, 23 June 2021, https://www.nytimes.
com/2021/06/23/magazine/exxon-mobil-engine-no-1-board.
html

38. ExxonMobil, 'Notice of 2021 Annual Meeting and Proxy
Statement', 16 March 2021, https://www.sec.gov/Archives/edgar
/data/34088/000119312521082140/d94159ddefc14a.htm. The two
members were Wan Zulkiflee and Jeff Ubben. Exxon also added a
third in Michael Angelakis, but the total number of seats only
increased by two because William Weldon was expected to retire
having reached the mandatory age limit.

39. Jennifer Hiller and Svea Herbst-Bayliss, 'Exxon, Activist Spends
Over $65 mln in Battle for Oil Giant's Future', Reuters, 15 April
2021, https://www.reuters.com/business/energy/exxon-activist-
spend-over-65-mln-battle-oil-giants-future-2021-04-15/

40. Aguirre, 'Little Hedge Fund Taking Down Big Oil'.

41. Leslie Kaufman and Saijel Kishan, 'Calstrs's Crucial Phone Call Eased Path for Activists' Exxon Win', *Bloomberg*, 18 June 2021, https://www.bloomberg.com/news/articles/2021-06-18/calstrs-s-crucial-phone-call-eased-path-for-activist-s-exxon-win?

42. Akshat Rathi and Kevin Crowley, 'Exxon Pushed for Net-Zero by Activist Shareholder', *Bloomberg*, 22 February 2021, https://www.bloomberg.com/news/articles/2021-02-22/exxon-pushed-by-activist-investor-to-set-net-zero-climate-goal

43. David G. Victor, 'Energy Transformations: Technology, Policy, Capital and the Murky Future of Oil and Gas', 3 March 2021. https://reenergizexom.com/documents/Energy-Transformations-Technology-Policy-Capital-and-the-Murky-Future-of-Oil-and-Gas-March-3-2021.pdf

44. 'Q1 2023 Quarterly Results', ExxonMobil, 28 April 2023, https://corporate.exxonmobil.com/-/media/Global/Files/investor-relations/annual-meeting-materials/proxy-materials/ExxonMobil-3_16_21-Shareholder-Letter.pdf

45. 'Letter to the Board of Directors', Reenergize Exxon, 22 February 2021, https://reenergizexom.com/materials/letter-to-the-board-of-directors-february-22/

46. Kevin Crowley and Akshat Rathi, 'Exxon's Plan for Surging Carbon Emissions Revealed in Leaked Documents', *Bloomberg*, 5 October 2020, https://www.bloomberg.com/news/articles/2020-10-05/exxon-carbon-emissions-and-climate-leaked-plans-reveal-rising-co2-output?sref=jjXJRDFv

47. Call transcript of ExxonMobil's 2021 Annual General Meeting held on 26 May.

48. 'Engine No. 1's Penner Accuses Exxon of Trying to Entrench Board', CNBC, 26 May 2021, https://www.cnbc.com/video/2021/05/26/engine-no-1s-penner-accuses-exxon-of-trying-to-entrench-board.html

49. Katherine Dunn and Sophie Mellor, 'ExxonMobil Faces Historic Loss in Proxy Shareholder Battle over Future of Its Board', *Fortune*, 26 May 2021, https://fortune.com/2021/05/26/exxonmobil-agm-landmark-vote-shareholders/

50. Kevin Crowley and Scott Deveau, 'Exxon CEO Is Dealt Stinging Setback at Hands of New Activist', *Bloomberg*, 26 May 2021, https://www.bloomberg.com/news/articles/2021-05-26/tiny-exxon-investor-notches-climate-win-with-two-board-seats

51. Rebecca Leber, 'Why Big Oil Should Be Worried after a Day of Reckoning', *Vox*, 27 May 2021, https://www.vox.com/22455347/exxon-board-shell-oil-news-chevron-engine-no-one

52. Andrew Ross Sorkin et al., 'Activist Investor Leads a Rebellion against ExxonMobil', *New York Times*, 27 May 2021, https://www.nytimes.com/2021/05/27/business/dealbook/exxon-mobil-engine-no-1.html

53. Jennifer Hiller and Svea Herbst-Bayliss, 'Engine No. 1 Extends Gains with a Third Seat on Exxon Board', Reuters, 2 June 2021, https://www.reuters.com/business/energy/engine-no-1-win-third-seat-exxon-board-based-preliminary-results-2021-06-02/

54. Hiller and Herbst-Bayliss, 'Little Engine No. 1 Beat Exxon with Just $12.5 mln – Sources'.

55. Zingales and McLean, 'Engine No. 1, David vs Exxon Goliath'.

56. Milton Friedman, 'A Friedman Doctrine – The Social Responsibility of Business Is to Increase Its Profits', *New York Times*, 13 September 1970, p. 17.

57. DealBook, 'A Free Market Manifesto That Changed the World, Reconsidered', *New York Times*, 11 September 2020, https://www.nytimes.com/2020/09/11/business/dealbook/milton-fried-man-doctrine-social-responsibility-of-business.html

58. Zingales and McLean, 'Engine No. 1, David vs Exxon Goliath'.

59. 'Business Roundtable Redefines the Purpose of a Corporation to Promote "An Economy That Serves All Americans"', Business Roundtable, 19 August 2019, https://www.businessroundtable.org/business-roundtable-redefines-the-purpose-of-a-corporation-to-promote-an-economy-that-serves-all-americans

Chapter 12: The Next Steps

1. 'Avoiding the Storm: Climate Change and the Financial System – Speech by Sarah Breeden Given at the Official Monetary & Financial Institutions Forum, London', Bank of England, 15 April 2019, https://www.bankofengland.co.uk/speech/2019/sarah-breeden-omfif

2. Karl Mathiesen, 'Leading Climate Lawyer Arrested after Gluing Herself to Shell Headquarters', *Climate Home News*, 16 April 2019, https://www.climatechangenews.com/2019/04/16/leading-climate-lawyer-arrested-gluing-shell-headquarters/

Index